Oscarina Batalha
Sonia Sena Alfaia
Frank George Cruz

Acid silage of pirarucu waste

Oscarina Batalha
Sonia Sena Alfaia
Frank George Cruz

Acid silage of pirarucu waste

In the feeding of light commercial layers

ScienciaScripts

Imprint

Cover image: www.ingimage.com

This book is a translation from the original published under ISBN 978-620-2-03210-0.

Publisher:
Sciencia Scripts
is a trademark of
Dodo Books Indian Ocean Ltd. and OmniScriptum S.R.L publishing group

120 High Road, East Finchley, London, N2 9ED, United Kingdom
Str. Armeneasca 28/1, office 1, Chisinau MD-2012, Republic of Moldova, Europe
Managing Directors: Ieva Konstantinova, Victoria Ursu
info@omniscriptum.com

Printed at: see last page
ISBN: 978-620-6-76073-3

Dedication,

To my parents Orestes and Natalina Batalha, who built a marvellous home for their children.

ACKNOWLEDGEMENTS

Firstly to God, for always guiding my steps.

To my family who have always been with me, for their companionship, love and affection.

To my supervisors, Dr Sonia Sena Alfaia, Dr Frank George Guimarães Cruz and Dr Rogério de Jesus, for the opportunity, trust, attention and guidance, which were fundamental at all times.

To FAPEAM-Fundação de Amparo à Pesquisa do Estado do Amazonas, for the grant awarded.

To the Postgraduate Programme in Agriculture in the Humid Tropics.

To my dear technician at COTI-INPA's Fish Technology Laboratory, Aparecida Bitencourt, for her help with the laboratory analyses.

To my dear friends Emanuel Leite and Lenoir Santos for their support at the start of the project.

To the good-hearted Emmaluelle Moraes, Filipe Leite and Vaneza de Sá, for their co-operation when I needed it most.

To the Frank Cruz family, Ana Paula Costa, André Silva, Brenna Freitas, Cristiane Guimarães, Francisco Chaves, Jadilson Barroncas, Julmar Feijó, João Paulo Rufino, Lucas Melo, Natalia Santos, Ramon Melo and Valcely Costa.

Thank you so much!

SUMMARY

CHAPTER 1

1. GENERAL INTRODUCTION

Today, poultry production, both broiler and layer, is recognised as an internationalised, uniform economic activity with no geographical borders. The traditional raw materials used in animal feed formulations are maize and soya meal, which account for around 70% of production costs. In this context, the search for alternative ingredients, especially those from energy and protein sources, to make up for the lack of such ingredients in seasonal periods is becoming more frequent.

From an environmental point of view, there is concern about the proper disposal of the solid waste generated daily through the processing of fish in production centres. According to Feltes et al. (2010) the waste generated by the fish agro-industry is mainly destined for animal feed because it is rich in proteins and omega-3 fatty acids.

Among the methods for using fish waste in animal feed, fish silage stands out because it offers economic advantages, as it requires simple technology and is independent of scale (Hisano et al. 2012), as well as using material with low operating costs to transform waste materials into products of high nutritional quality that can minimise problems with environmental pollution, as well as serving as an ingredient in feed formulation.

The Bacalhau da Amazônia project has led to the construction of a fish warehouse in the municipality of Maraã - Amazonas, specialising in the production of dried salted fish, mainly using the pirarucu species (Arapaima gigas). Currently, a large part of the pirarucu from the Mamirauá Sustainable Development Reserve is processed at the Maraã Fish Processing Unit.

In 2014, 2,490 units were processed at the Maraã Fish Processing Unit, generating a total of approximately 155 tonnes of fresh pirarucu, of which approximately 30% was treated as waste. As a way of utilising fish waste, the production of silage has emerged as an alternative for transforming pirarucu waste into excellent sources of protein with high biological properties for inclusion in monogastric animal feed.

This study is divided into two chapters. The first looks at the physico-chemical characteristics and digestibility of piraruc waste acid silage flour in rations for light commercial layers; and the second looks at the performance, quality and sensory analysis of eggs from light commercial layers fed pirarucu waste acid silage flour.

1. LITERATURE REVIEW

2.1 Pirarucu (Arapaima gigas, Schinz 1822)

Arapaima gigas, popularly known as pirarucu (figure 1), are large fish that have a wide distribution in the Amazon basin, their exclusive habitat. It belongs to the order: Osteoglossiforme (bony tongue), family: Arapaimatidae, genus: Arapaima, species: Arapaima gigas, Schinz 1822.

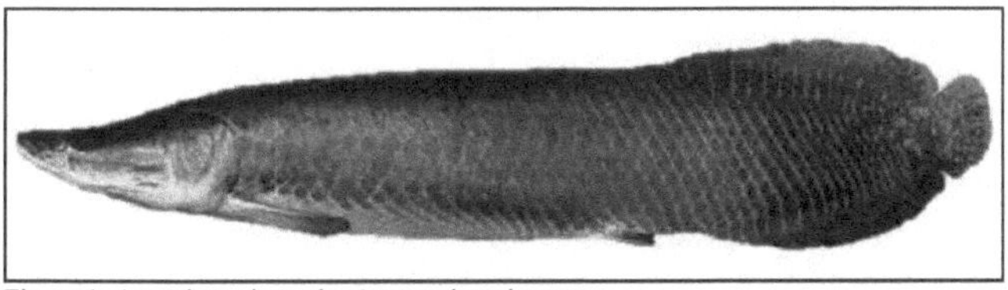

Figure 1. Arapaima gigas, the Amazonian pirarucu.

It is characterised by its flattened and ossified head, long and wide body with thick cycloidal scales and a bony tongue used as sandpaper (Oliveira 2007). The dorsal and caudal fins are located at the posterior end of the body. Its colouration is dark with red areas, especially in the tail region (Mueller and Green 2005). It has two respiratory systems, consisting of gills (aquatic respiration) and a swim bladder modified to function as lungs in aerial respiration, which makes it emerge every 10-20 minutes. As an adult, it can measure between two and three metres in length and weigh 200 kg (Ono et al. 2004).

Their reproductive behaviour involves forming monogamous couples, parental care, building nests on the substrate of lakes and ponds and they spawn in parcels (Cavero and Fonseca 2016). Spawning in the Central Amazon is more concentrated in the months of October to March and in the Western Amazon, due to the difference in water regime, spawning is concentrated from January to May, coinciding with the start of the rains (Ono et al. 2004).

Other important zootechnical characteristics that draw attention to this species are: easy adaptation to the consumption of food and balanced commercial rations, high quality lean meat free of intramuscular spines, high fillet yield, high market demand and, when raised in captivity, a fattening rate of 10kg in one year (Ono and Kehdi 2013).

Various studies have been carried out on their biology (Galvão de Lima and Batista 2012), reproduction in captivity (Pedroza Filho et al. 2016) and nutrition (Rodrigues et al. 2015) in order to understand their characteristics in farming systems.

1.1.2 Production and commercialisation

According to the FAO (2016), the estimated world production of fish in 2015 was 158 million tonnes, of which 136.2 million tonnes were used as food by people around the world.

With regard to world production, from both extractive fishing and aquaculture, Brazil contributed 1,240,813 tonnes, ranking 18th overall and 17th in the world aquaculture ranking, with production of 415,649 tonnes in 2009 (Brasil 2010). Brazil has 3,000 species of fish, a large number of which have potential for use in fish farming, such as dourado, jaú, matrinxã, piau, pintado, pirarucu, jandiá and tambaqui (Brasil 2014).

The northern region of Brazil led continental extractive fishing in 2011, with production of 137,144.5 tonnes, which represented 55% of national production. Although fishing production in the state of Amazonas fell by around 10% compared to 2010, it remained the most significant, with 63,743.3 tonnes, representing 40.3% of the total caught, followed by the states of Pará (55,402.7 tonnes) and Maranhão with 25,743.5 tonnes, respectively (Brasil 2013).

Pirarucu has been exploited in the Amazon basin since the 18th century, when it was commercialised, and still is today, in dried salted slices to be used as a substitute for cod (Mueller and Green 2005). Its sale in salted form was abundant near the main centres of consumption, such as Manaus/AM and Santarém/PA, until the 1960s and in Belém/PA, some small companies and establishments sold dried-salted pirarucu (Imbiriba 2001).

Due to intense commercial exploitation, the pirarucu population began to decline. Faced with this problem, IBAMA, through Ordinance 039 of 02/12/1987, established a minimum catch size of 1.50 metres, which allows individuals to reach sexual maturity and reproductive conditions. Through

Ordinance 480 of 15/03/1991, the species is prohibited from being caught from 1 December to 31 May in the entire Amazon region, the Araguaia River and the Tocantins River.

The pirarucu fishery under management began in 2002 and has grown exponentially over 11 years, from a production of 45,542 tonnes to 1,266,457 tonnes in 2013 and 1,656 tonnes in 2015. Also in terms of the fishing environment, it went from three areas and three municipalities in 2002 to 19 catching areas located in 13 municipalities (Sepror 2015).

Cavero and Fonseca (2016) argue that both captive pirarucu and pirarucu from managed areas compete for the same market, the difference being in price. According to the authors, pirarucu sold in the form of cigars (without the head and guts) is sold for R$3.50/kg. The fish that arrives in Manaus (fillet with skin and ribs) is sold for between R$8.00/kg and R$14.00/kg, resulting in the price of the fillet in natura between R$9.00/kg and R$15.50/kg.

1.1.3 Pirarucu yield

Carcass yield is one of the key factors in defining the profitability of processing (Honorato et al. 2014). Pirarucus ranging in size from 71 to 190 cm achieve an average fillet yield of 57%; the head represents 12 to 13.5%; the scales 6.6 to 9.7%; the viscera 5.4 to 6.0% and the vertebral column 14.2 to 18% of the animal's live weight (Oliveira 2007).

According to Fogaça et al. (2011) in an evaluation of the performance of pirarucu farmed for fillet production in different classes, they found that fish from 7.0 to 9.0 kg (group I); 11.0 to 13.0 kg (group II) and 14.0 to 16.0 kg (group III), showed 47.43%; 48.62% and 49.79% in fillet yield.

In addition to the meat, other projects can be developed to use the waste left over from the industrial processing of pirarucu. The leather represents 7 to 9 % of the animal's weight and can be used to make shoes, belts and bags; the carcasses can be used as regional artefacts or to make silage, which is an alternative product to be used in animal feed (Onu et al. 2004; Lima et al. 2014).

2.2 Fish silage

2.2.1 History

In the year 525 BC, the Romans used fish waste to transform what looks like silage today into a thick sauce known as garum, which was prepared with one or more species of fish, always with the gills, entrails, salt and various spices, which were placed in closed ceramic pots to undergo the autolysis process (Maia Junior and Sales 2013).

Fish silage is not a new product. The method for making it originated in Finland in the 1920s and was adapted from methods of preserving fodder by adding formic, hydrochloric and sulphuric acids and other ingredients such as molasses (Hisano and Borghesi 2011).

In the 1940s, other countries such as Canada, Australia, Norway and Germany began to produce fish waste silage in small quantities (Santos 2000). It was only in Denmark, Poland and Norway, from the 1960s onwards, that production became effective on a commercial scale, using the method of combining acids to produce feed for poultry and pigs or incorporated into rations as a protein supplement, used to feed domestic animals and aquaculture fish (Ferraz de Arruda 2004).

Therefore, despite much research into the production and use of silage, unlike in other countries, fish silage is not commercially produced by the Brazilian industry, and the reason why it is not commonly used is related to the inconsistent supply of quality (Ferraz de Arruda et al. 2006).

2.2.2 Principles and methods of making fish silage

According to FAO (2012), the principle for making silage consists of properly grinding fish waste in a hammer mill equipped with a sieve containing holes with diameters of ten millimetres. Disposable waste is used (viscera, scales, skeletons and heads), resulting from the processing of marine and freshwater fish generated in local commercialisation (Nascimento et al. 2014).

The process for making fish silage is simple, practical and economical, as it does not require sophisticated equipment and high-cost procedures, such as those used to produce fishmeal (Oetterer 2002). It is a viable alternative for production in small quantities and with low energy costs (Vidotti et al. 2002).

It consists of evenly acidifying the pH of the ground mass to prevent untreated particles from rotting. Periodic stirring is necessary to favour the liquefaction of the tissues. The silage must remain in a pH range between 3.9 and 4.5 to avoid the action of pathogenic microorganisms, at a maximum temperature of 30 °C and a minimum of 20 °C, as liquefaction occurs slowly below this level (Vidotti and Gonçalves 2006).

Two methods are used to make fish silage: by adding inorganic or organic acids, such as acetic, sulphuric, hydrochloric, formic and propionic - called acid silage (Abimorad et al. 2009; Carmo et al. 2008; Fogaça 2008; Sales and Oliveira 2014) or using lactic acid-producing microorganisms together with carbohydrate sources, resulting in biological silage (Gonçalves and Viegas 2007; Yamamoto et al. 2007; Godoy et al. 2008; Honorato et al. 2011).

2.3 Organic fish silage

There are various fermentation methods used to produce fish silage, one of which uses bacteria from the lactic acid group: Lactobacillus sp and Lactococcus sp or yeasts from the genera Hansenula and Saccharomyces, giving rise to biological silage (Machado 2010). In this case, carbohydrate sources such as sugar cane molasses and microorganism sources such as expired yoghurt and whey can be used, always using fungicides and antioxidants to preserve the quality of the dough (Vidotti and Gonçalves 2006).

According to Machado (2010), temperature is a determining factor when it comes to carrying out the fermentation process, as it determines the generation of a stable product. For this, the ideal temperature is 40°C for 48 hours, with a pH between 3.5 and 4.0.

In this regard, Nascimento et al. (2014) compared three silages (acid, biological and enzymatic) and observed that the acidity standards were more evident in the acid silage (4.08 to 3.82) and the highest in the enzymatic silage (7.13 to 5.14), while the biological silage was intermediate (6.84 to 4.48); on the other hand, when checking the temperature, the statistical values were less evident (29.98 and 29.33) for the biological and enzymatic silages.

Mousavi et al. (2013) analysed the pH of three silage production methods (mineral acid, organic acid and biological) for use in shrimp feed. The authors concluded that the organic silage produced with beet pulp (a carbohydrate source) reached a stable level after 20 days, with a pH of 4.00, which they described as the ideal method because there was no need for neutralisation.

Table 1 shows the centesimal composition of biological fish silages produced with biological inoculum and carbohydrate sources and their respective authors. Table 1. Centesimal composition, pH and temperature of biological silage.

Inoculo	T°C	pH	DM (%)	PB (%)	EE (%)	CZ (%)	Authors
Molasses and *Lactobacillus plantarum*	28,1	4,43	33,05	33,62	25,61	13,58	Honorato *et al.* 2012
Sugar e yoghurt	28,0	4,5	97,5	29,19	8,29	29,83	Nascimento *etal.* 2014
Cabbage and papaya e wheat flour	30,0	4,0	85,66	38,94	4,77	31,98	Oliveira *et al.* 2012
Molasses and *Lactobacillus* sp	38,0	4,4	34,5	39,9	14,5	18,0	Ramírez Ramírez et al. 2013

pH: Hydrogen Potential; T °C: Temperature; DM: Dry Matter, CP: Crude Protein, EE: Ethereal Extract and CZ: Ash.

2.4 Acid fish silage

Acid or chemical fish silage is a form of acid hydrolysate, which is presented as a liquefied product. The principle consists of acidifying the pH by adding acids that will prevent bacteriological putrefaction of the ensiled material, while the enzymes present in the fish meat break down the proteins and liquefy them (Carmo et al. 2008).

During the process, the ensiled material needs to be homogenised daily to facilitate the incorporation of the acids and promote the enzymatic action naturally present in the muscles, and consequently the proteins will be hydrolysed, making the nitrogen more soluble, which can be seen in the increase in the levels of short-chain amino acids and peptides (Vidotti 2001).

The addition of organic (formic and propionic) and inorganic (hydrochloric, sulphuric, nitric and phosphoric) acids in the preparation of fish silage causes the proteolytic enzymes responsible for protein and lipid autolysis present in fish flesh to accelerate their activity, so that pH values are in the 3.9 to 4.5 range (Vidotti and Gonsalves 2006; Benites and Souza-Soares 2010).

According to Vidotti and Gonçalves (2006) silage produced by adding acidifiers remains aseptic because it does not give off unpleasant odours, and does not present problems with insects and pathogenic microorganisms such as salmonella.

2.4.2 Organic acids in silage production

According to Brazilian legislation, through Decree No. 55.871 of 26/06/1965, organic acids are classified as preservatives or acidulents, defined as substances that are added to food with the aim of delaying

microbial or enzymatic action, protecting the food against spoilage (Brasil 2001).

Due to their low solubility, intense flavour and low toxicity to the human body, short-chain organic acids, characterised by their carboxyl functional group (COOH) (Table 2), have been the most widely used in animal feed for some years (Soccol 2002, Viola and Vieira 2007). The choice of acidifiers depends on cost, availability and bactericidal action (Benites and Souza-Soares 2010).

Table 2. Chemical properties of different organic acids.

Names	Formula	Shape	Solubility in H2O
Formic	HCOOH	Liquid	Very good
Acetic	CH3COOH	Liquid	Very good
Propionic	CH3CH2COOH	Liquid	Very good
Lactic	$CH_3CH(OH)COOH$	Liquid	Good
Fumaric	COOHCH:CHCOOH	Solid	Low
Citrus	$COOHCH_2C(OH)(COOH)CH_2COOH$	Solid	Good

Source: Bellaver and Scheuermann (2004).

According to Sucasas (2011), the choice of acidifying ingredient is very important and must take into account the risks to the environment and human health, as well as costs and accessibility, as some acids are controlled by the Federal Police, Civil Police or the Brazilian Army. The same author presents a list of government-controlled acids (Table 3) and a list of the factors that make acids suitable for use in silage production.

Table 3. Relationship between acids and degree of risk and accessibility.

Acids	Degree of risk to the environment	Degree of health risk (*)	Controlled Product
Acetic	2	3	Yes-PF
Citric Anhydrous	2	1	No
Formic	2	3	Yes-PF
Phosphoric	3	2	No
Muri Atticus	3	3	No
Propionic	2	3	No
Sulphuric	4	4	Yes-PF

PF: Federal Police
*1: Small risk, 2: Moderate risk, 3: Severe risk, 4: Extreme risk.
Source: Sucasas (2011).

According to Brasil (2003), the chemical products controlled by the Federal Police are listed in Lists I, II, III, IV and their respective Addendums in Annex I, in accordance with Ordinance 1.274 of 25 August 2003. According to the Ordinance, control is only applied to consumption of more than two litres per month, and small-scale production is exempt.

The combination of weak acids, propionic and formic, produces less acidic silage, as well as fungicidal activity, eliminating the need for final neutralisation (Sucasas 2011).

The activity of organic acids in microorganisms is related to the reduction in pH and the ability of their carboxyls to dissociate, i.e. in their undissociated form, the acids passively penetrate the microbial cell, where they release protons and anions, resulting in a reduction in intracellular pH, inhibiting the action of enzymes and leading to the death of the microorganisms (Viola and Vieira 2007).

2.4.2 Nutritional composition of acid silage

Despite the changes that take place during the production process, silage produced from waste or whole fish retains similar chemical and nutritional characteristics to the original material, but these depend on the species, head size, rearing system, fat and other factors that make up the silage (Vidotti 2011; Cândido 2016).Vidotti et al. (2003) assessed the protein composition of silage made from three types of waste material. They observed that acid silage produced from whole fish from salt water had a higher crude protein content (69.91%) compared to silage produced from freshwater fish (44.38%). Table 4 shows the centesimal composition of chemical fish silage produced with formic and propionic acids, and their respective authors.

Table 4. Centesimal composition and pH of chemical silage made with different acids.

Material Used	Acids	PH	DM (%)	PB (%)	EE (%)	CZ (%)	Authors
Tambaqui guts	Formic and Propionic (1:1) 3% v/p	3,0 a 3,75	54,92 ± 2,24	10,71 ± 0,75	77,38 ± 2,57	1,72 ± 0,04	Borghesi *etal.* 2013
Shrimp waste	Formic and Propionic (1:1) 3% v/p	4,0	20,20	66,14	13,86	14,06	Costa 2007
Tilapia waste	Phosphoric (2%) and acetic acid (3%)	3,02 a 3,88	100	30,63	47,89	14,12	Fernandes *et al.* 2007
Pirarucu guts	Formic and Propionic (1:1) 3% v/w	3,02	40,78± 1,15	17,38± 2,06	68,97± 3,76	2,69±0, 01	Lima et al. 2013

DM: Dry Matter, CP: Crude Protein, EE: Ethereal Extract, CZ: Ash, pH: Hydrogen Potential.

One of the factors that can influence the quality of the nutritional value of fish silage is prolonged storage, as a large number of free amino acids are present in the silage due to the activity of endogenous enzymes, which can result in a reduction in nutritional value (Borghesi 2004).

However, the nutritional value of fish silage can be increased by limiting protein hydrolysis, as short-chain peptides (di- and tripeptides) are better and more absorbed by fish than free amino acids (Honorato et al. 2012).

Another factor considered relevant to the quality of silage is lipid oxidation, which can lead to the formation of peroxides that complex with proteins through covalent bonds, resulting in the destruction of amino acids (Ferraz de Arruda 2004). The same author argues that centrifugation is the best method for lipid extraction, making it possible to increase storage time and adapt feed formulations.

Ge Fei et al. (2010) evaluated the basic chemical composition, amino acids, trace elements, degree of hydrolysis and protein content in silage prepared with or without heating and stored for 40 days at different temperatures. The authors concluded that storage conditions, temperature and time contribute to the degree of hydrolysis and protein content in silage prepared with fish waste.

2.5 Digestibility of fish silage

High digestibility is related to the nutritional quality of fish silage, as well as the integral presence of the amino acids that make up the fish (Ferraz de Arruda et al. 2006). According to Borguesi et al. (2013), drying silage together with soya meal, in a 1:1 ratio, showed better results in terms of performance and digestibility in experiments with fish.

Yamamoto et al. (2007) evaluated the performance and digestibility of nutrients in lambs fed diets containing 8% fish waste silage, and found that lambs fed diets formulated with the inclusion of freshwater fish silage showed greater digestion of ether extract (3.20 g/kg/day) and a better coefficient of digestibility of ether extract (90.39%) compared to the control diet (2.25%) for this study.

Honorato et al. (2012) aimed to assess the apparent digestibility of dry matter, protein and crude energy of fish silage at different times (7, 15, 30 days) of ensiling for feeding pacu (Piaractus mesopotamicus), observing that there was no significant difference in apparent digestibility as a function of time.

In a study carried out by Guzel et al. (2011) to assess the growth of rainbow trout (Oncorhynchus mykiss) with inclusion levels of 25, 50 and 100 per cent, they found that at the end of the experiment, the group of fish fed diets formulated with 50 per cent silage showed significantly higher weight gain compared to the other formulations.

In a study with Santa Inês sheep, using rations with inclusion levels of 25, 50 and 75%, Santos (2010) concluded that shrimp waste can replace the protein source from soya by up to 75%. Fish waste silage has great potential to be fed to small ruminants, as it improves meat quality and carcass shape, as well as making up for the lack of protein in certain foods where there is a shortage of this nutrient (Rahmi et al. 2008).

2.5.1 Effect of fish silage on poultry feed

Liquid fish silage can be added directly to the diet or dried for use as a feed ingredient

(Goddard and Perret 2005).

Al-Marzooqi et al. (2010) evaluated the effect of four levels (0, 10, 20 and 30 per cent) of sardine silage on the performance and quality of meat characteristics of broiler chickens reared in closed and open side housing systems and found that chickens fed diets containing 10 and 20 per cent gained more weight than chickens fed 30 per cent fish silage. They concluded that fish silage can replace up to 20% of soya meal without affecting growth performance or the sensory quality of the meat.

In a study to assess the blood biochemical composition and meat quality of broiler chickens, under the effect of supplementation with fish waste (acid silage and surimi waste powder), Darsana and Sreekumar (2012) concluded that total substitution of fish waste fulfils the nutritional requirements without causing toxicity, in addition to maintaining meat quality.

Kjos et al. (2001) concluded that the inclusion of 50 g kg^{-1} of diet containing 12% total protein did not affect feed intake, egg production, yolk fatty acid composition, yolk colour or egg sensory quality compared to the control in laying hen diets.

Ramirez-Ramirez et al. (2013) evaluated inclusion levels (0, 10, 20, 30 per cent) on the performance and meat quality of 160 Japanese quails (Coturnix coturnix japonica). They observed that the levels did not affect growth rate and feed conversion, nor did they provide significant effects on carcass yield and sensory quality of the meat in the different treatments.

Ponce and Gernat (2002) evaluated the replacement of soya and maize meal with different levels of tilapia meal in broiler chicken feed. The results showed that animals fed 10, 20 and 30 per cent increased weight gain, consumption and feed conversion at 14 to 28 days of age.

In an evaluation of the effect of five levels (0%, 2%, 4%, 6% and 8%) of inclusion of tilapia (Oreochromis niloticus) waste meal, Eyng et al. (2010) concluded that the addition of 8% did not affect the yield and sensory quality of broilers in the one to 42 day phase.

Fish waste can therefore be used to produce various products, one of which is silage, which has excellent protein value and can replace fishmeal and soya in animal diets (Maia Junior and Sales

2013).

3. REFERENCES

Abimorad, E. G.; Strada, W. L.; Schalch, S. H. C.; Garcia, F.; Castellani, D.; Manzatto, M. R. 2009. Fish silage in artisanal feed for nile tilapia. Pesquisa Agropecuária Brasileira, 44: 519-525.

Al-Marzooqi, W.; Al-Farsi M.A.; Kadim, I.T.; Mahgoub, O.; Goddard, J. S.; 2010. The effect of feeding different levels of sardine fish silage on broiler performance, meat quality and sensory characteristics under closed and open-sided. Asian-Australasian Journal Animal Sciences, 23: 16141625.

Amâncio, A. L. L.; Silva, J. H. V; Lima, C. B.; Araújo, J. A.; Coelho, M. S.; Oliveira, E. R. A.; Lima, M. R.; Silva, N. V.; Rodrigues, M. L. 2010. Nutritional value of fish silage and its use in animal feed. Centro de Pesquisa e Gestão de Recursos Pesqueiros do Litoral Norte, Boletim Técnico-Científico Cepnor, 10: 79-93.

Brazil, Ministry of Health. 2001. Health Surveillance Secretariat. Decree No. 55871 of 26 March 1965. Modifies Decree No. 50.040 of 24 January 1961 regulating the use of food additives, amended by Decree No. 691 of 13 March 1962. (www.planalto.gov.br/decreto). Accessed on 12/10/2015.

Brazil, 2003. Ordinance No. 1.274, of 25 August 2003. Subjects to control and inspection the chemical products listed in Lists I, II, III, IV and their respective Addenda. Official Gazette, Brasília, DF. Section 1, Extra Edition 164: 1-2. (www.pf.gov.br/servicos-pf/produtos-quimicos/legislacao/portaria1274.pd). Accessed on 12/10/2015.

Brazil. 2011. Fisheries and aquaculture statistical bulletin. Ministry of Fisheries and Aquaculture-MPA, Brasília, (www.icmbio.gov.br). Accessed on 03/05/2016.

Brazil. 2012. Fisheries and aquaculture statistical bulletin, Ministry of Fisheries and Aquaculture-MPA, Brasilia.

Brazil 2012. Statistical bulletin of fisheries and aquaculture - Brazil. Brasilia: Ministry of Fisheries and Aquaculture, 2013.

Brazil. 2014. 1st Brazilian Fishing and Aquaculture Yearbook. Ministry of Fisheries and Aquaculture-MPA p.1-136. (formsus.datasus.gov.br/newimgarq/.pdf).

Bellaver, C.; Scheuermann, G. 2004. Organic acid linkages in poultry production. In: Avisui Conference, 3, Florianópolis, Santa Catarina.

Benites, C.I.; Souza-Soares, L.A. 2010. Fish waste silage meal co-dried with rice bran: a viable alternative. Archivos de Zootecnia, 59:447-450.

Borghesi, R. 2004. Physico-chemical, nutritional and biological evaluation of acid, biological and enzymatic silages made with waste and residues from the processing of Nile tilapia (Oreochromis niloticus). Master's dissertation, Luiz de Queiroz College of Agriculture, Piracicaba, State of São Paulo, 96p.

Borghesi, R.; Ferraz de Arruda, L.; Oetterer, M. 2007. Fish silage in the feeding of aquatic organisms. Bulletin of the Food Processing and Research Centre, Curitiba, 25: 329-339.

Borghesi, R.; Lima, L. K. F.; Sucasas, L. F. A.; Marto, V. C. O.; Oetterer, M. 2013. Preparation of acid and acid co-dried silages from tambaqui (Colossoma macropomum) viscera. Technical Circular, 106. Corumbá, Embrapa Pantanal. 6p. (www.cpap.embrapa.br/publicacoes/online/CT106.pdf). Accessed on 04/05/2016.

Cavero, B. A. S.; Fonseca, F. A. L. 2016. Pirarucu: current situation and prospects in the Amazon region. Panorama da Aquicultura. (www.panoramadaaquicultura.com.br). Accessed on 03/05/2016.

Carmo, J. R.; Pimenta, J. C.; Pimenta, M. E. S. G.; Oliveira, M. M.; Logato, P. V. R.; Ferreira, L. O. 2008. Characterisation of acid silages from tilapia (Oreochromis niloticus) waste. Nutritime Electronic Journal, 5: 664-672.

Cândido, R. S. 2016. Digestibility of fish silage with microbiological inoculum for growing pigs. Master's thesis, Federal University of Ceará, Fortaleza, 44p.

Costa, C. N. 2007. Evaluation of acid silage from white shrimp waste (Litopenaeus vannamei) as a protein source for feeding Nile tilapia (Oreochromis niloticus). Master's thesis, Federal University of Recôncavo da Bahia, Cruz das Almas, Bahia, 45p.

Darsana, M. G.; Sreekumar, K. P. 2012. Effect of processed fish wastes supplementation on blood biochemical and meat composition of broiler chicken. Iranian Journal of Veterinary Research, Shiraz University, 13: 186-194.

Eyng, C., Nunes, R. V.; Pozza, P. C.; Silva, W. T. M.; Navarini, F. C.; Henz, J. R. Flour from tilapia filleting industry waste in broiler rations. Revista Brasileira de Zootecnia, 39:2670-2675.

Feltes, M. M. C.; Correia, J. F. G.; Beirão, L. H.; Block, J. M.; Ninow, J. L.; Spille, V. R. 2010. Alternatives for adding value to fish industrialisation waste. Brazilian Journal of Agricultural and Environmental Engineering,14: 669-677.

Ferraz de Arruda, L. 2004. Utilisation of Nile tilapia (Oreochromis niloticus) to obtain silage and oil as by-products. Master's dissertation, Escola Superior de Agricultura "Luiz de Queiroz", Universidade de São Paulo, Piracicaba, State of São Paulo, 78p.

Ferraz de Arruda, L.; Borghesi, R.; Brum, A.; Regitano D'Arce, M.; Oetterer, M. 2006. Nutritional aspects of nile tilapia (Oreochromis niloticus) silage. Food Science and Technology, 26: 749-753.

Fernandes, J. B. K.; Bueno, R. J.; Rodrigues, L. A.; Fabregat, T. El H. P.; Sakomura, N. K. 2007. Acid silage from tilapia filleting waste in feed for juvenile piauçu (Leporinus macrocephalus). Acta Scietiarum Animal Sciences, 29:339-344.

Fogaça, F. H. S. 2008. Technical Recommendations for the Preparation of Marine Shrimp Silage. Technical Circular 49, Teresina, Embrapa Meio-Norte, 4p

Fogaça, F. H. S.; Oliveira, E. G.; Carvalho, S. E. Q.; Santos, F. J. S. 2011. Yield and composition of pirarucu fillet in different weight classes. Acta Scientiarum Animal Sciences, 33: 95-99.

FAO. 2012. The State of World Fisheries and Aquaculture. Food and Agriculture Organisation of United Nations, Rome, 209 p.

FAO, Food and Agriculture Organisation of the United Nations. 2014. The State of the World's Fisheries and Aquaculture (www.fao.org.br). Accessed on 03/05/2016.

Ge Fei; Gui Lin; Tao Yu-Gui; Huang Yin. 2010. Effects of Storage Conditions on Quality of Fish Silage Prepared from Channel Catfish. Food Science, 31: 290-293.

Gonçalves, L. U.; Viegas, E. M. M. 2007. Production, characterisation and biological evaluation of shrimp waste silage for nile tilapia. Arquivo Brasileiro de Medicina Veterinária e Zootecnia, 59:1021-1028.

Godoy, H. B. R.; Landell Filho, L. C.; Bianchini Sobrinho, E. M.; Godoy, M. 2008. The use of fish filleting by-product silage in the feeding of growing pigs serum parameters. Brazilian Journal of Veterinary Research and Animal Science, São Paulo, 45: 429-436.

Guzel, S.; Yazlak, H.; Gullu, K.; Ozturk, E. 2011.The effect of feed made from fish processing waste silage on the growth of rainbow trout (Oncorhynchus mykiss). African Journal of Biotechnology, 10: 5053-5058.

Hisano, H.; Borghesi, R. 2011. Preparation of acid silage from surubim (Pseudoplatystoma sp.) viscera. Technical Circular 18, Embrapa Agropecuária Oeste, Dourados, Minas Gerais. 4p.

Hisano, H.; Ishikawa, M. M.; Portz, L. 2012. Production of acid silage from surubim (Pseudoplatystoma sp.) viscera and evaluation of digestibility for nile tilapia. Revista Brasileira de Saúde e Produção Animal, Salvador, 13: 872-879.

Honorato, C. A.; Stech, M. R.; Carneiro, D. J. 2011. Organic fish waste silage in diets for Nile tilapia fry. Revista Acadêmica de Ciências Agrárias Ambiente, Curitiba, 9: 371-377.

Honorato, C.; Frizzas Ganga, O.; Carneiro, D. J. 2012. Digestibility of fish silage with different storage times for feeding pacu (piaractus mesopotamicus). Ensaios e Ciência: Ciências Biológicas, Agrárias e da Saúde, Campo Grande, 16: 85-95.

Honorato, C. A.; Caneppele, A.; Matoso, J. C.; Prado, M. R.; Siqueira, M. S.; Souza, L. R. O. 2014. Physical characterisation of Surubim (Pseudoplatystoma sp.) and pirarucu (Arapaima gigas) fillets. Arquivos de Ciências Veterinárias e Zoologia da Unipar, Umuarama, 17: 237-242.

Imbiriba, E. P. 2001. Potential for breeding pirarucu, Arapaima gigas, in captivity. Acta Amazônica, 31: 299-316.

IBAMA-Brazilian Institute for the Environment and Renewable Natural Resources. 2009. (www.ibama.gov.br). Accessed on 02/05/2016.

Kjos, N. P.; Herstard, O.; Skrede, A.; Overland, M. 2001. Effects of dietary silage and fish facto n performance and egg quality of laying herns. Canadian Journal of Animal Science, 1432: 245-251.

Galvão de Lima, L.; Batista, V. S. 2012. Ethnoichthyological studies on the pirarucu Arapaima gigas in Central Amazonia. Acta Amazonica, 42: 337-344.

Lima, J. 2014. Brazil invests in research to increase the productivity of pirarucu in captivity. Science and Culture, 66: 12-13.

Machado, T. M. 2010. Biological fish silage.Panorama da aquicultura. 8: 3032.

Maia Junior, W. M. M.; Sales, R. O. 2013. Functional properties of obtaining acid and biological silage from fish waste. Revista Brasileira de Higiene e Sanidade Animal, 07: 126-156.

Mousavi, S. L.; Mohammadi, G.; Khodadadi, M.; Keysami, M. A. 2013. Silage production from fish waste in cannery factories of Bushehr city using mineral acid, organic acid, and biological method. Intl J Agri Crop Sci. 6: 610-616.

Mueller, O.; Green, A. 2005 *Arapaima gigas - Market study:* current status of arapaima global trade and perspectives on the Swiss, French and UK markets, Bio Trade Facilitation Programme-UNCTAD.

Nascimento, M. S.; Freitas, K. F. S.; Silva, M. V. 2014. Production and characterisation of fish waste silage sold in the public market of Parnaíba - PI. Enciclopédia Biosfera. 10: 450.

Oetterer, M. 2002. Industrialisation of farmed fish. Guaíba: Agropecuária, 200p.

Oliveira, P. R. 2007. Quality of Pirarucu (Arapaima gigas, Shinz 1822) from fish farms, stored on ice, frozen and its derived products. PhD Thesis, National Research Institute of Amazonas/Federal University of Amazonas, Manaus, Amazonas. 130p.

Oliveira, C. R. C. Nutritional evaluation of silage meal in broiler diets. 2012. Dissertation (Master's in Animal Science) - Federal Rural University of Pernambuco, Department of Animal Science, Recife, 82p.

Ono, E. A.; Halverson, M. R.; Kubitza, F. 2004. Pirarucu: no fish converts better. Panorama of Aquaculture, 14: 1-25.

Ono, E. A.; Kehdi, J. 2013. Manual of Good Production Practices for Pirarucu in Captivity. Sebrae, Brasília. 46 p.

Pedroza Filho, M. X.; Munoz, A. E. P.; Rodrigues, A. P. O.; Rezende, F. P., Lima, A. F.; Mataveli, M. 2016. Overview of the pirarucu production chain. Boletim Ativos da Aquicultura. (https://www.embrapa.br/panorama-da-cadeia-produtiva-do-pirarucu). Accessed on 03/03/2017.

Ponce, L. E.; Gernat, A. G. 2002. The effect of using different levels of tilapia by- product meal in broiler diets. Poultry Science, 81:1045-1049.

Rahmi, M.; Faid, M.; Elyachioui, M.; Berny E. H.; Fakir, M.; Ouhssine, M. 2008. Protein rich ingredients from fish waste for sheep feeding. African Journal of Microbiology Research, 2: 073-077.

Ramírez-Ramírez, J. C.; Ibarra, J. I.; Romero, F. A.; Ulloa, P. R.; Ulloa J. A.; Shirai, K. M.; Cordoba, B. V.; Manzano, M. A. M. 2013. Preparation of Biological Fish Silage and its Effect on the Performance and Meat Quality Characteristics of Quails (Coturnix coturnix japonica). Brazilian Archives of Biology and Technology, 56:1002-1010.

Rodrigue, A. P. O.; Moro, G. V.; Santos, V. R. V. 2015. Feeding and nutrition of pirarucu (Arapaima gigas). Embrapa Fisheries and Aquaculture, 24 p.

Santana-Delgado, H.; Avila, E.; Sotelo, A. 2008. Preparation of silage from Spanish mackerel (Scomberomorus maculatus) and its evaluation in broiler diets. Animal Feed Science and Technology, 141: 129-140.

Santos, N. F. 2000. Processing, chemical and nutritional characterisation of biological fish waste silage for use in animal feed. Master's dissertation, Federal University of Ceará, Fortaleza, Ceará. 84p.

Santos, M. M. 2010. Performance of Santa Inês sheep fed levels of protein from shrimp waste as a substitute for a conventional protein source. Master's dissertation. São Cristóvão: UFS. 34 p.

Sales, R. O.; Oliveira, A. C. 2014. Influence of Nile tilapia (Oreochromis niloticus, L.) harvest silage in rations for pigs in the growth and finishing phases with rats (wistar) at growth. Revista Brasileira de Higiene e Sanidade Animal, 8:19-36.

Sepror-Secretary of Rural Production of Amazonas. 2015. Fishing production. (www.sepror.am.gov.br.). Accessed on 07/03/2017.

Soccol, M. C. H. 2002. Optimising the shelf life of farmed tilapia (Oreochromis niloticus), minimally processed and stored under refrigeration. Master's dissertation, Escola Superior de Agricultura "Luiz de Queiroz"/Universidade de São Paulo, Piracicaba, State of São Paulo. 124p.

Sucasas, L. F. A. 2011. Evaluation of fish processing waste for the development of co-products aimed at increasing sustainability in the production chain. PhD Thesis, University of São Paulo, Piracicaba, São Paulo. 167p.

Vidotti, R. M. 2001, Production and utilisation of fish silage in the nutrition of pacu (P iaractus mesopotamicus). PhD Thesis, Jaboticabal: UNESP/CAUNESP. 65p.

Vidotti, R. M.; Carneiro, D. J.; Viegas, E. 2002. Growth rate of pacu, Piaractus mesopotamicus, fingerlings fed diets containing co-dried fish silage as replacement of fish meal. Journal of Applied Aquaculture, 12: 77-88.

Vidotti, R. M.; Viegas, E. M. M.; Carneiro, D. J. C. 2003. Amino acid composition of processed fish silage using different raw materials. Animal Feed Science and Technology, 105: 199-204.

Vidotti, R. M.; Gonçalves, G. S. 2006. Production and characterisation of tilapia silage, meal and oil and their use in animal feed. Instituto de Pesca. (www.pesca.sp.gov.br). Accessed on 25/06/2015.

Viola, E. S.; Vieira, S. L. 2007. Supplementation of organic acidifiers in broiler diets: zootechnical performance and intestinal morphology. Revista Brasileira de Zootecnia, 36:1097-1104.

Vidotti, R.M.; Gonçalves, G. S. 2011. Production and Characterisation of Silage, Flour and Oil from Tilapia and their Use in Animal Feed. Instituto de Pesca. (www.pesca.sp.gov.br). Accessed on 25/06/2015.

Vidotti; R. M. 2011. Technologies for fully utilising fish. Technical Course in Intensive Fish Farming (www.pesca.sp.gov.br). Accessed on 25/06/2015.

Yamamoto, S. M.; Silva Sobrinho, A. G.; Vidotti, R. M.; Homem Junior, A. C.; Pinheiro, R. S. B.; Buzzulini, C. 2007. Performance and nutrient digestibility in lambs fed diets containing fish waste silage. Revista Brasileira de Zootecnia, 36:1131-1139.

4. OBJECTIVES

4.1 General

- To evaluate the effect of acid silage produced from the residual biomass of pirarucu Arapaima gigas (Schinz 1822) in diets for light layers.

4.2 Specific

- Applying the acid method to crushed pirarucu waste to obtain silage flour as an ingredient for commercial laying diets;
- Characterise the chemical, nutritional and organoleptic composition of acid silage and flour produced from residual pirarucu biomass;
- Determine the digestibility of pirarucu residue acid silage meal in rations for light commercial layers
- To evaluate the performance of light layers fed with levels of inclusion of acid silage flour from pirarucu waste and the quality and sensory analysis of eggs.

CHAPTER 2

PHYSICOCHEMICAL CHARACTERISTICS AND DIGESTIBILITY OF PIRARUCU WASTE ACID SILAGE MEAL IN RATIONS FOR LIGHT COMMERCIAL LAYERS

Physico-chemical composition and digestibility of pirarucu waste acid silage meal in rations for light commercial layers

SUMMARY

The aim of this study was to evaluate the physico-chemical and nutritional characteristics of pirarucu waste acid silage meal in rations for light commercial layers, in terms of apparent digestibility and energy metabolisation. Seventy-two 71-week-old Hisex White layers were used in a completely randomised design with six replicates of six birds each. The treatments consisted of control (feed without product inclusion) and experimental feed (with 3% acid silage inclusion). The ensiled biomass was light brown in colour; had a slightly acidic aroma; a slightly creamy consistency; a pH of 4.38±0.11; with values of 84.16% dry matter; 40.06% crude protein; 26.82% ether extract; 9.31% mineral matter, 65.16 g/kg calcium and 22.90 g/kg phosphorus. Significant differences were observed in the digestibility of crude protein, non-nitrogenous extract (soluble carbohydrates), ether extract, mineral matter, metabolisable energy and the metabolisation coefficient of metabolisable energy. The results of this study indicate that acid silage produced from residual pirarucu biomass can be included in the form of meal at a level of 3% in rations for light layers, showing good nutrient digestibility and the potential to be used as an energy source.

KEYWORDS: alternative feed, silage, metabolisable energy, nutrients, fish waste

1. INTRODUCTION

The Amazon is home to a great diversity of aquatic organisms, which are the main source of income and subsistence for riverside communities that live off the exploitation of fishing resources along the Solimões-Amazonas river and its tributaries (Costa et al. 2013). Among the species most exploited and consumed by the regional population is the pirarucu (Arapaima gigas, Schinz 1822), known as the cod of the Amazon, whose meat is highly valued on the market, especially when it is commercialised using the dry-salted procedure.

Only the fish fillets are used for processing pirarucu, in the form of frozen fillets or salted-dried fish, while the waste (carcasses, viscera, fins, scales and skins) is discarded in the environment. However, some of these by-products have the potential to be utilised as new products.

In this context, one product that has been highlighted in recent studies carried out in Brazil to utilise fish waste is silage. This is a viable alternative for using filleting waste (Costa et al. 2009), as well as being a simple technique with low operating costs used to transform waste materials into products of high nutritional quality that can minimise problems with environmental pollution, as well as serving as an ingredient in feed formulation.

As for the use of these fish waste silages in animal feed, they have already shown the potential to provide economic benefits to poultry production systems, since their feed represents 70% of total production costs, where maize and soya are the main ingredients used as energy and protein sources in feed formulations. In this sense, the use of this waste generated by the fish agro-industry to transform by-products into alternative foods offers accessibility advantages, especially in regions of Brazil that face obstacles with grain logistics and high-cost raw materials (Cruz et al. 2016).

From this perspective, the use of silage as an alternative ingredient aims to mainly the productive and economic aspect, without harming, above all, the animal's performance and physiology. Therefore, it is necessary to have a clear understanding of the protein and mineral composition, energy concentration, accessibility of nutrients, as well as their utilisation by the birds' bodies (Oliveira et al. 2012).

In view of the above, this research was carried out with the aim of evaluating the chemical and nutritional characteristics of acid silage from pirarucu waste and the use of its meal in rations for light commercial layers on apparent digestibility and energy metabolisation.

2. MATERIAL AND METHODS

2.1 Location of the study area

The experiment was carried out in two stages: the first was at the Fish Technology Laboratory of the Technology and Innovation Coordination Office (CITO) of the National Institute for Amazonian Research (INPA), with the procedures for obtaining acid silage from pirarucu (Arapaima gigas, Schinz 1822) waste and its technical and nutritional characterisation.

The second stage was carried out at the Federal University of Amazonas - UFAM, in the Poultry sector of the Faculty of Agricultural Sciences, located in the southern sector of the University Campus in Manaus-AM, with geographical coordinates of latitude 3° 06' 14" S, longitude 59° 58' 46" W and an altitude of 260 metres. According to the classification proposed by Koppen, the region's climate is of the Af type (hot and humid), with an average temperature ranging from 27 to 29°C, average annual rainfall of 2,286 mm and relative humidity of 80%.

2.2 Conducting the Experiment

2.2.1 Raw materials

The raw material used came from the Mamirauá Sustainable Development Reserve, Amazonas-Brazil, and was acquired in February 2016 from a fish processing warehouse, consisting of the carcass (backbone and ribs). This waste was transported in sealed raffia bags to the COTI-INPA Fish Technology Laboratory.

2.2.2 Sour silage production

The waste was shredded in a Multinox electric shredder and divided into plastic buckets with a capacity of 15 litres, where each container received 10 kg of shredded mass. For every 10 kg, a mixture of propionic and formic acids was added in a 1:1 ratio, at 3% of the volume of the acid solution in relation to the weight of the waste under constant homogenisation, following the methodology proposed by Borghesi et al. (2008) (Figure 1). The containers were kept at room temperature for three days.

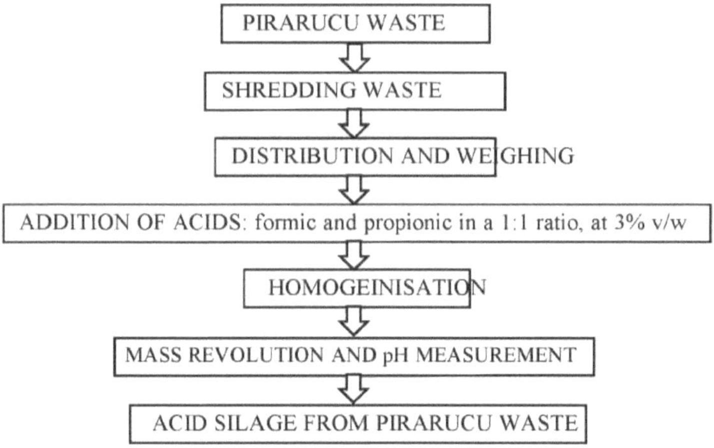

Figure 1. Flowchart for obtaining acid silage.

In order to achieve complete uniformity, the silage material was turned every 24 hours during the three days of storage to allow greater contact between the acids and the residues and guarantee the quality of the silage, from which samples were taken for pH measurement. The pH was determined using an OHAUS pH meter (Starter 3100), with results expressed to two decimal places.

Changes in the product were monitored by observing the organoleptic characteristics, based on colour, aroma and consistency. At the end of the three days, to reduce the humidity, the ensiled material was placed in aluminium trays and taken to a ventilation oven at 65 °C for 72 hours. During this time, the silage was turned over to ensure uniform drying and to obtain a quality dry product.

2.3 Chemical composition

The dry matter (DM), crude protein (CP), ether extract (EE), mineral matter (MM) and hydrogen potential (pH) values were determined at INPA's Food Chemistry and Physical Chemistry Laboratory, following the methodology used by Silva and Queiroz (2012). The moisture content of the sample was determined by the gravimetric method in an oven at 105°C until constant weight. The crude protein content was obtained using the Kjeldahl method, the ether extract was realised after extraction with hexane using the Soxhlet method and the mineral matter content was measured by incineration in a muffle furnace at 550°C for 4 hours. Total carbohydrates (TC) were obtained using the equation: 100 - (% Moisture + %PB + %MM + %EE). At INPA's Thematic Soil and Plant Chemistry Laboratory, the minerals Calcium (Ca) and Phosphorus (P) were digested in nitro-perchloric extract according to the methodology described by Sarruge and Haag (1974), and the concentrations quantified by atomic absorption spectrophotometer (Ca) and by colorimetry (P) in the spectrophotometer using ammonium molybdate and ascorbic acid.

After determining the chemical composition, the dried product was packaged and sent to the poultry sector at UFAM, where it was ground in a grinder using a sieve with a mesh size of 0.10 mm to obtain acid silage flour from pirarucu waste.

2.4 Metabolisation experiment in light layers

The rations (Table 1) were formulated using the computer software Supercrac (2004) to meet the nutritional requirements of the animals and in accordance with the ingredient values provided by the Brazilian Tables for Poultry and Pigs (Rostagno et al. 2011), with the exception of the composition of the pirarucu residue acid silage meal.

Table 1. Composition of control experimental rations and rations containing 3 % acidic pirarucu waste silage.

Ingredients	Experimental Rations	
	0,0%	3,0%
Maize 7.88%	62,255	61,756
Soya bran 46%	25,871	23,758
Sour silage	0,000	3,000
Calcitic limestone	9,239	8,806
Bicalcium phosphate	1,693	1,716
Premix Vit. Min.[1]	0,500	0,500
Salt	0,350	0,350
DL- Methionine 99	0,092	0,114
Total (Kg)	100,000	100,000
Calculated nutrients		
Metabolisable energy, kcal-1/kg	2692,59	2629,18
Crude protein, %	17,000	17,00
Methionine + Cystine, %	0,627	0,620
Methionine, %	0,360	0,404
Calcium, %	4,000	4,000
Available phosphorus, %	0,400	0,400
Sodium, %	0,157	0,156

[1] Guarantee levels per kilogram of product: Vitamin A 2,000,000 IU, Vitamin D3 400,000 IU, Vitamin E 2,400 mg, Vitamin K3 400 mg, Vitamin B1 100 mg, Vitamin B2 760 mg, Vitamin B6 100 mg, Vitamin B12 2.400 mcg, Niacin 5,000 mg,

Calcium Pantothenate 2000 mg, Folic Acid 50 mg, Coccidiostat 12,000 mg, Choline 50,000 mg, Copper 1,200 mg, Iron 6,000 mg, Manganese 14,000 mg, Zinc 10,000 mg, Iodine 100 mg. Selenium 40 mg. Vehicle Q.S.P. 1.000 g.

In the poultry sector's feed mill, the feed ingredients were weighed before mixing began. The ingredients were then mixed for 15 minutes in a vertical mixer with a capacity of 500kg.

72 layers of the Hisex White strain were used at 71 weeks of age.

The birds were housed in cages measuring 1.0m long, 0.45m deep and 0.45m high with internal partitions measuring 0.50m along the length. The birds were weighed in order to standardise the plots, with an average weight of 1.50±0.0029 kg and distributed in a completely randomised experimental design (DIC), where the treatments consisted of a control diet (based on corn and soybean meal) and an experimental diet with a 3% inclusion of acidic pirarucu waste silage flour in the feed, with six replicates of six birds each. The experimental period lasted 12 days, including seven days for the birds to adapt to the diets and facilities and a further five days for collecting excreta and data according to the methodology proposed by Rodrigues et al. (2005) and Sakomura and Rostagno (2007).

The experiment was carried out in a shed with a fibre cement roof measuring 17.0m long, 3.5m wide and 3.20m high, with a central aisle and a row of wire cages on each side, trough-type feeders and niplle-type drinkers.

The digestibility of the nutrients in the feed was determined using the total excreta collection method. For this procedure, trays were used which were attached to the floor of the cages and lined with plastic, from which the excreta was collected twice a day, in the early morning (8am) and late afternoon (4pm), and then packed in airtight bags labelled according to treatment and stored in a freezer.

At the end of the excreta collection period, the samples were thawed at room temperature, homogenised per experimental unit and a composite sample was taken for drying in a forced ventilation oven at 55°C for 72 hours, then they were ground and together with the experimental rations they were analysed to determine the levels of dry matter, crude protein, ether extract and crude fibre and ash, according to the techniques described by Silva and Queiroz (2012).

After the analyses, the nutrient digestibility coefficients, metabolisable energy values and apparent crude energy metabolisation coefficients of the feeds were determined according to the equations described by Matterson et al. (1965) for evaluating feeds (Sakomura & Rostagno, 2007). The equations used were:

CDA = (Nutrient consumed - Faecal nutrient) / (Nutrient consumed) x 100;

MRE = (Gross energy ingested - Gross energy excreted) / (Dry matter ingested);

CMAEB = (Apparent Metabolisable Energy / Gross Energy) x 100

In which:

ADC = Apparent digestibility coefficient

AME = Apparent metabolisable energy

CMAEB = Coefficient of apparent metabolisation of gross energy

2.5 Statistical analyses

The data collected, with the exception of the pH results, was subjected to analysis of variance (ANOVA), followed by the Tukey mean comparison test ($P<0.05$), using the Statistical Analysis System - SAS (2008) computer programme.

3. RESULTS AND DISCUSSION

The results of pH development over a 72-hour period for acid silage and fresh pirarucu waste are shown in Table 2. The pH was monitored during the period in which the biomass remained ensiled. Initially, it was 6.61±0.01 in the crushed waste, which decreased with the addition of the acid mixture to 4.22±0.22 on the first day and 4.38±0.11 on the last day. The final pH result obtained in this study is within the 4.5 limit recommended by Benites and Souza-Soares (2010) to prevent microbiological action in acidic fish silage.

Table 2. pH values of pirarucu waste and acid silage*.

Days/hours	Waste in natura	Sour silage
Home	6,61±0,01	-

1st (24 hours)	-	4,22±0,22
2nd (48 hours)	-	4,28±0,19
3rd (72 hours)	-	4,38±0,11

* Averages taken from three repetitions ±standard deviation.

The initial pH of the in natura residue is within the range of values (6.67±0.01 to 6.81±0.01) observed by Oliveira et al. (2014) during the 6 to 12 day ice storage period of pirarucu muscles, proving that the residue presented favourable conditions for use in making acid silage.

With a final pH value of 4.38±0.11, no changes were observed that could jeopardise the quality of the silage during the experimental period. Acidic pH produces conditions that reduce or prevent the growth of undesirable bacteria that cause protein decomposition and putrefaction of the silage mass (Vieira et al. 2015).

The combination of organic acids such as formic and propionic, in a ratio of (1:1) or (3:1) with 3% of the volume by weight, has been recommended in the preparation of acid silage, as they maintain the bacteriostatic properties at a higher pH, not requiring the need for neutralisation, due to their power to stabilise the pH of the biomass (Maia Junior and Sales; 2013). In this study, it was not necessary to neutralise the silage before drying it for inclusion in the feed, as the pH was close to the recommended limit for microbiological prevention.

Using the visual method during ensiling, it was observed that the colour characteristics of the residual mass changed with the addition of the mixture of formic and propionic acids, changing from the slightly pinkish colour natural to fish in natura to a light grey. The contact of the acid with the biomass causes reactions that release the nutrients present in it (Tanuja et al. 2014), inhibiting the growth of pathogenic and spoiling microorganisms (Venturoso et al. 2016), thus blocking the bad odour (Vieira et al. 2015). Therefore, the peculiar aroma of pirarucu disappeared, giving way to an acidic smell, which became milder as the days went by, indicating that the added acids had been volatised, making the product and work environment free of undesirable insects.

The liquid formed was incorporated into the biomass using the same process to allow maximum contact between the acids and the residual fragments, giving it a creamy consistency. This result is in line with the visual observations made by Vasconcelos et al. (2011) who reported that the homogeneous mass began to liquefy within 24 hours and increased until the end of the experiment.

The formation of liquids in fish silage is attributed to the continuous hydrolysis of proteins due to the action of proteolytic enzymes, naturally present in fish muscle (Ribeiro et al. 2015), which are accelerated by the addition of acids and reach their highest activity at pH values between 2 and 4 (Tomczak-Wandzel and Mędrzycka 2013). In general, the rate of liquefaction depends on the type of raw material, its freshness, enzymatic activity, the physiological state of the fish and other factors (Al-Abri et al. 2014). In this study, the formation of liquid was not very pronounced due to the structure of the waste, which was not ground but only fragmented. The time the material spent in the silage process also contributed to this factor.

During the 72-hour storage period, the ensiled biomass took on a light brown colour, with a slightly acidified aroma and a slightly creamy, not very liquid consistency, corroborating those observed by Vasconcelos et al. (2011) in a study on the physicochemical standards and yield of acid silage from Nile tilapia *(Oreochromis niloticus).*

The final aroma, which was slightly acidified, was de-characterised during the drying and grinding of the product for inclusion in the feed, showing a characteristic fishy smell. This is due to the use of acids, which are less corrosive and easier to handle (Ferraz de Arruda *et al.* 2007).

The results of the chemical composition of acid silage and *fresh* pirarucu waste are shown in Table 3. The chemical composition of the acid silage of pirarucu waste showed significant differences ($P<0.05$) compared to the *in natura* waste, except for calcium and phosphorus.

Table 3. Chemical composition of pirarucu waste and acid silage flour.

Chemical composition	Waste	Sour silage	P Value	CV (%)
Humidity, %	68,58	15,84	-	-
Dry Matter, %	$31{,}42^{b}$	$84{,}16^{a}$	0,01*	0,65
Crude Protein, %	$15{,}71^{b}$	$40{,}05^{a}$	0,01*	3,04
Ether extract, %	$6{,}46^{b}$	$26{,}81^{a}$	0,01*	1,13
Mineral Matter, %	$3{,}42^{b}$	$9{,}31^{a}$	0,01*	5,93
Total carbohydrates %	$5{,}82^{b}$	$7{,}97^{a}$	0,01*	1,63
Calcium, g/Kg	66,32	65,16	0.58^{ns}	3,70
Phosphorus, g/Kg	27,09	22,90	0.34^{ns}	4,54

CV - Coefficient of variation; * Averages followed by lowercase letters on the line differ by Tukey's test at 5% significance ($P <0.05$); ns - not significant.

The chemical composition of the in natura pirarucu residue found in the present study showed only crude protein results below those obtained by Oliveira et al. (2014) who found 79.51±0.52% moisture, 17.56±0.12% CP; 0.62±0.02% EE; 0.87±0.06% CZ and 1.44% total carbohydrates in the dorsal part and 77.88±0.53% moisture, 16.10±0.37% CP; 2.49±0.03% EE; 0.84±0.05% CZ and 2.69% carbohydrates in the ventral part of the pirarucu muscle from fish farms.

This difference in chemical composition between acid silage and the fresh waste from pirarucu is related to the removal of moisture, which allowed the nutrients to be concentrated. However, the acids used in silage help enzymes to hydrolyse the proteins into peptides and amino acids and the lipids present in the waste (Tomczak-Wandzel and Mçdrzycka 2013). This accelerated breakdown by the incorporation of acids, as well as reducing the pH, also breaks down bones and cartilage, preventing the growth of spoilage bacteria (Goddard and Perret 2005).

The crude protein concentration found in this study was higher than the 38.12% obtained by Ferraz de Arruda et al. (2009), using sulphuric and formic acid treatment (3:1) in silage produced with fresh waste and similar to the 40.62±0.12, 40.45±0.06 and 40.38±0.06% crude protein recorded by Ramasubburayan et al. (2013) using 2, 2.5 and 3% formic acid in the preparation of acidic fish silage. According to the aforementioned authors, the reduction in crude protein in silage made with acids is related to continued protein hydrolysis.

The mineral matter levels (9.31%) found in this study were higher than those observed by Boscolo et al. (2010) in the order of 6.57, 5.91 and 5.77 at 7, 91 and 201 days of storage of acid silage from tilapia waste, where he states that the amount of mineral matter found in fish silage is due to its residual constitution, where the minerals are more concentrated in this waste.

The ether extract value of 26.82% observed in the acid silage of pirarucu waste is associated with the ventral part, referring to the ribs, where the fish's higher fat content is concentrated and which contributed to the greater release of fat during the ensiling period. Tanuja et al. (2014) obtained 39.19±0.53% ether extract from acid silage of carp viscera waste, and this value presented by the author shows that, like the ventral part, the viscera are also the main site of lipid deposition in fish.

The minerals analysed, calcium and phosphorus, found in the acid silage are due to the use of the parts of the bone structure that make up the pirarucu, such as the ribs and vertebrae, which can influence the results. The calcium and phosphorus contents obtained in the present study were well above those found by Hisano et al. (2012), whose values were 0.06% calcium and 0.27% phosphorus, and those obtained by Dale and Valenzuela (2016) in the order of 1.01% calcium and 1.08% phosphorus in the mineral composition of dried salmon acid silage.

The chemical composition of fish silage can vary according to the raw material used in production, such as: species of fish, portion within the same species, rearing system, stage of development of the fish, part analysed and type of acids used to facilitate hydrolysis (Borghesi et al. 2007; Oliveira et al. 2012).

The results observed for the apparent digestibility coefficients of the nutrients in the feed are shown in Table 4. Significant differences were observed (P<0.05) in the digestibility coefficients of the non-nitrogenous extract, ether extract and mineral matter. It was observed that layers fed a ration containing 3% acid silage meal showed better utilisation of ether extract and mineral matter compared to the control ration. These results may be related to the higher content of these components in the acid silage meal made from pirarucu waste.

Table 4. Apparent digestibility coefficients of the reference ration and the experimental ration containing 3% pirarucu waste acid silage flour for light layers.

Coefficients of Digestibility (%)	Experimental Rations 0,0%	3,0%	P Value	CV (%)
Dry Matter	70,62	70,87	0.89^{ns}	2,94
Crude Protein	$55{,}48^{a}$	$35{,}91^{b}$	0,01*	9,87
Crude fibre	4,99	4,64	0.32^{ns}	13,45
Non-nitrogenated extract	$83{,}61^{b}$	$88{,}67^{a}$	0,01*	1,20
Ethereal extract	$67{,}59^{b}$	$74{,}36^{a}$	0,02*	3,51
Mineral Matter	$14{,}92^{b}$	$32{,}22^{a}$	0,03*	20,54

CV - Coefficient of variation; * Averages followed by lowercase letters on the line differ by Tukey's test at 5% significance (P <0.05); ns - not significant.

In this study, the birds used to analyse digestibility were 71 weeks old, which may have directly influenced nutrient digestibility. According to Arruda et al. (2012) and Fernandes et al. (2015), older birds have a fully developed digestive system with greater production of digestive enzymes, which allows for a greater possibility of permanence and absorption of nutrients contained in food.

Previously, layers fed diets containing a 3 per cent inclusion of acidic pirarucu waste silage showed lower crude protein digestibility ($P<0.05$) compared to those fed the reference diet. This result may be attributable to the small fragments of peptides and free amino acids in hydrolysates, which are absorbed more quickly than the amino acids in intact proteins, making them easily available for energy production rather than for protein synthesis (Hernández et al. 2013).

These results disagree with those obtained by Al-Marzooqi et al. (2010), who found a higher digestibility coefficient for amino acids in sardine silage made with hydrochloric acid, concluding that proteins in hydrolysed form are easily used in protein synthesis by birds.

There were no differences ($P>0.05$) in the digestibility of dry matter and crude fibre between the experimental feed containing the 3% inclusion of acidic pirarucu waste silage and the reference feed, as the layers had similar performance.

Several studies have demonstrated the benefits of silage made with the addition of organic acids to the feed, such as that carried out by Widjastuti et al. (2011) who observed better results in birds fed diets containing 4% silage made with the addition of 3% formic and propionic acid (1:1), and Rahman and Koh (2016) concluded that shrimp meal treated with 3% formic acid provided better digestibility of nutrients in broiler diets.

The results of the apparent metabolisable energy and the apparent metabolisable energy metabolisation coefficient of the feed are shown in Table 5. The values of the apparent metabolisable energy coefficients and apparent metabolisable energy metabolisation coefficients showed significant differences ($P<0.05$) for layers fed the ration containing 3% acid silage, due to the greater use of energy sources in the diet.

Table 5: Apparent metabolisable energy (aME) and apparent gross energy metabolisation coefficient (aGEMC) of the control ration and the experimental ration (containing 3% pirarucu residue acid silage flour - FSARP) for light layers.

Digestibility	Experimental Rations	P Value	CV (%)

coefficients (%)	Reference feed	Feed with 3% FSARP		
EMa[1]	2921,18b	3253,01[a]	0,01*	0,79
CMEMB2	78,93b	81,14[a]	0,01*	0,23

CV - Coefficient of variation; * Averages followed by lowercase letters on the line differ by Tukey's test at 5% significance ($P < 0.05$); ns - not significant.

[1] Apparent Metabolisable Energy

[2] Coefficient of Apparent Metabolisation of Crude Energy

In a study carried out to obtain apparent metabolisable energy values, Oliveira et al. (2014) found energy values ranging from 3,804 kcal/kg to 3,842 kcal/kg in rations made with fish silage ensiled with different carbohydrate sources in broiler diets. In this study, the apparent metabolisable energy of the feed containing acidic pirarucu waste silage meal was 3253.01 kcal/kg, which is close to the values reported in the literature.

However, the difference in energy values between the studies may be related to variations in the chemical composition of the feed (Calderano et al. 2010), and its better utilisation by the birds. In this way, the nutritional evaluation of alternative feedstuffs for egg production represents an investigation of great technical and scientific value, as in many cases it is not fully utilised, given the scarcity of regional diversity (Melo et al. 2015).

4. CONCLUSION

The results of this study indicate that the acid silage produced from residual pirarucu biomass can be included in the form of flour at a level of 3% in rations for light layers, showing good nutrient digestibility and the potential to be used as an energy source.

5. REFERENCES

Al-Marzooqi, W.; Al-Farsi M.A.; Kadim, I.T.; Mahgoub, O.; Goddard, J. S.; 2010. The effect of feeding different levels of sardine fish silage on broiler performance, meat quality and sensory characteristics under closed and open-sided. Asian-Australasian Journal Animal Sciences, 23: 1614-1625.

Al-Abri, A. S.; Mahgoub, O.; Kadim, I. T.; Al-Marzooqi, W.; Goddard, S. J.; Al-Farsi, M. 2014. Processing and evaluation of nutritive value of fish silage for feeding Omani sheep. Journal of Applied Animal Research, 42: 406-413.

Arruda, A. M. V.; Melo, A. S.; Oliveira, V. R. M.; Souza, D. H.; Oliveira, J. F. 2012. Nutritional evaluation of cassava manioc hay with free-range birds. Acta Veterinaria Brasilica, 6: 204-210.

Benites, C.I.; Souza-Soares, L.A. 2010. Fish waste silage meal co-dried with rice bran: a viable alternative. Archivos Zootecnia, 59: 447-450.

Borghesi, R., Ferraz de Arruda, L.; Oetterer, M. 2007. Fish silage in the feeding of aquatic organisms. Bulletin of the Food Processing and Research Centre, Curitiba, 25: 329-339.

Boscolo, W. R.; Santos, A. M.; Buzanello, M. C. V.; Feiden, A., Bittencourt, F.; Signor, A. A. 2010. Microbiological and bromatological evaluation of acid silage obtained from Nile tilapia (Oreochromis niloticus) filleting industry waste. Ciências Agrárias, Londrina, 31: 515-522.

Borghesi, R., Portz, L., Oetterer, M. & Cyrino, J. E. P. 2008. Apperent digestibility coeffcient of protein and amino acids of acid, biological and enzymatic silage for Nile tilapia (Oreochromis niloticus). Aquaculture Nutrition, 14: 242-248.

Calderano, A. A.; Gomes, P. C.; Albino, L. F. T.; Rostagno, H. S.; Souza, R. M.; Mello, H. H. de C. 2010. Chemical and energy composition of plant-based feedstuffs determined in birds of different ages. Revista Brasileira de Zootecnia, 39: 320-326.

Costa, C. N.; Portz, L.; Hisano, H.; Druzian, J. I.; Ledo, C. A. da S. (2009). Acid silage from Litopenaeus vannamei shrimp waste in feed for Nile tilapia. Acta Scientiarum. Animal Sciences, Maringá, 31 (2),161-167.

Costa, T. V.; Silva, R. R. S.; Souza, J. L.; Batalha, O. S.; Hoshiba, M. A. 2013. Aspects of fish consumption and trade in Parintins. Boletim do Instituto de Pesca, São Paulo, 39: 63-75.

Cruz, F. G. G.; Rufino, J. P. F.; Melo, R. D.; Feijó, J. da C., Damasceno, J. L.; Costa, A. P. G. C. 2016. Socioeconomic profile of viticulture in the primary sector of the state of Amazonas, Brazil. Revista em Agronegócio e Meio Ambiente, Maringá, 9: 371-391.

Dale, N.; Valenzuela, C. 2016. Nutritional properties of dried salmon silage for broiler feeding. Animal Science Journal, 87: 791-795.

Ferraz de Arruda, L.; Borghesi, R.; Oetterer, M. 2007. Use of fish waste as silage: a review. Brazilian Archives of Biology and Technology, 50: 879-886,

Ferraz de Arruda, L.; Borghesi, R.; Portz, L.; Cyrino, J. E. P.; Oetterer, M. 2009. Fish Silage in Black Bass (Micropterus Salmoides) Feed as an Alternative to Fish Meal. Brazilian Archives of Biology and Technology, 52: 1261-1266.

Feijó, J. da C.; Cruz, F. G. G.; Melo, R. D.; Rufino, J. P. F.; Damasceno, J. L.; Costa, A. P. G. C.; Negreiros, T. de J. N. (2016). Yam flour (Dioscorea trifida L.) on the performance, egg quality and serum of light commercial layers. Revista Brasileira de Saúde e Produção Animal, Salvador, 17: 413-423.

Goddard, J.S.; Perret, J. S. M. 2005. Co-drying fish silage for use in aquafeeds. Animal Feed Science and Technology, 118: 337-342.

Hernández, C.; Olvera-Novoa, M. A.; Voltolina, D.; Hardy, R. W.; González- Rodriguez, B.; Dominguez-Jimenez, P.; Valverde-Romero, M.; Agramon-Romero, S. 2013. Use of tuna industry waste in diets for Nile tilapia, Oreochromis niloticus, fingerlings: effect on digestibility and growth performance. Latin American Journal of Aquatic Research, 41: 468-478.

Honorato, C.; Frizzas, G. O.; Carneiro, D. J. 2012. Digestibility of fish silage with different storage times for feeding pacu (Piaractus mesopotamicus). Ensaios e Ciência: Ciências Biológicas, Agrárias e da Saúde, Campo Grande, 16: 85-95.

Hisano, H.; Ishikawa, M. M.; Portz, L. 2012. Production of acid silage from surubim (Pseudoplatystoma sp.) viscera and evaluation of digestibility for nile tilapia. Revista Brasileira de

Saúde e Produção Animal, Salvador, 13: 872-879.

Fernandes, R. T. V.; Arruda, A. M. V.; Araújo, M. S.; Melo, A. da S.; Marinho, J. B. M.; Vasconcelos, N. V. B.; Lopes, F. de F.; Holanda, J. S. 2015. Energy values and digestibility coefficients of a traditional feed for Label Rouge birds at different ages. Acta Veterinaria Brasilica, 9: 108-113.

Maia Junior, W. M.; Sales, R. de O. 2013. Functional Properties of Acid and Biological Silage Obtained from Fish Waste. A Review. Brazilian Journal of Animal Hygiene and Health, 7: 126-156.

Matterson, L. D., Potter, L. M., Stutz, M. W., Singsen, E. P. The metabolisable energy of feed ingredients for chickens. Storrs: The University of Connecticut, Agricultural Experiment Station, 1965. 11p. (Research Report, 7).

Melo, R. D.; Cruz, F. G. G.; Feijó, J. da C.; Rufino, J. P. F.; Damasceno, J. L.; Dias, E. C. de S.; Brandão, A. B. T. 2015. Digestibility of Diets Containing Cara Flour (Dioscorea trifida L.) For Laying Hens. International Journal of Poultry Science, 14: 156-160.

Oliveira, D. D.; Pinheiro, J. W.; Fonseca, N. A. N.; Oba, A. 2012. Performance of broilers fed sunflower cake. Ciências Agrárias, Londrina, 33: 1979-1990.

Oliveira, C.R.C.; Ludke, M.C.M.M.; Ludke, J.V.; Lopes, E.C.; Pereira, P.S.; Cunha, G.T.G. 2014. Physicochemical composition and energy values of fish silage meal for broilers. Arquivo Brasileiro de Medicina Veterinária e Zootecnia, Belo Horizonte, 66: 933-939.

Oliveira, P. R.; Jesus, R. S. de; Batista, G. M.; Lessi, E. 2014. Sensory, physico-chemical and microbiological evaluation of pirarucu (Arapaima gigas, Schinz 1822) during storage on ice. Brazilian Jounal of Food Technology, Campinas, 17: 67-74.

Ramasubburayan, R.; Iyapparaj, P.; Subhashini, K. J.; Chandran, M. N., Palavesam, A.; Immanue, G. 2013. Characterisation and Nutritional Quality of Formic Acid Silage Developed from Marine Fishery Waste and their Potential Utilization as Feed Stuff for Common Carp Cyprinus carpio Fingerlings. Turkish Journal of Fisheries and Aquatic Sciences 13: 281-289.

Rahman, M.; Koh, K. 2016. Effects of Formic Acid-Treated Shrimp Meal on Growth Performance and Nutrient Digestibility in Broilers. Japan Poultry Science Association, 53: 208-212.

Ribeiro, I. de A.; Ribeiro, S. da C.A.; Castro, J. S. O.; Medeiros, G. K. C. Q. 2015. Obtaining and characterising flour from acid silage of farmed tambaqui filleting waste. Enciclopédia Biosfera, Centro Científico Conhecer - Goiânia, 11: 2015-2304.

Rodrigues, P.B.; Martinez, R.S.; Freitas, R.T.F.; Bertechini, A.G.; Filho, E.T. 2005. Influence of Collection Time and Methodologies on the Digestibility and Energy Value of Poultry Feed. Brazilian Journal of Animal Science, 34: 882-889.

Rostagno, H.S. Albino, L. F. T.; Donzele, J. L.; Gomes, P. C.; Oliveira, R. F.; Lopes, D. C.; Ferreira, A. S.; Barreto, S. L. de T.; Euclides, R. F. 2011. Brazilian tables for poultry and pigs: feed composition and nutritional requirements. Viçosa: Federal University of Viçosa, 252p.

Sakomura, N.K.; Rostagno, H.S. 2007. Research methods in monogastric nutrition. Jaboticabal: Funep, 283p.

Sarruge, J. R.; Haag, H. P. 1974. Chemical analyses in plants. Piracicaba: ESALQ. 56 pp.

Silva, D.J.; Queiroz, A.C. 2012. Food analysis: chemical and biological methods. 3ªedition, Viçosa: Federal University of Viçosa, 235p.

Supercrac. Minimum cost ration. Version 1.02, for Windows: TD Software, 2004.

Statistical Analysis System - SAS. SAS/STAT Software Version 9.2. Cary: SAS Institute Inc., 2008.

Tanuja S., Prafulla K. M., Kumar, A., Moharana, A. & Sujit, K. N. 2014. Shelf Life Study of Acid

Added Silage Produced from Fresh Water Fish Dressing Waste with and without the Addition of Antioxidants. International Journal of Agriculture and Food Science Technology, 5: 91-98.

Tomczak-Wandzel, R.; Mędrzycka, K. 2013. Preparation, composition and properties of fish silage produced with post-coagulation sludge. Environment Protection Engineering, 12: 80-233.

Vasconcelos, M. M. M.; Mesquita, M. do S. C. de; Albuquerque, S. P. 2011. Physico-chemical standards and yield of tilapia acid silage. Revista Brasileira Engenharia Pesca, 6: 27-37.

Venturoso, O. J., Reinicke, F., Silva, C. C., Vieira, E. O., Porto, M. O., Cavali, J.; Tortato, N.; Ferreira, E. 2016. Acid silage from fish waste for broiler chickens. Acta Veterinaria Brasilica, 10: 284-289.

Vieira, E. O., Venturoso, O. J., Reinicke, F., Silva, C. C.; Porto, M. O., Cavali, J., Vieira, N. T.; Ferreira, E. 2015. Production, Conservation and Health Assessment of Acid Silage Vicera of Freshwater Fish as a Component of Animal Feed. International Journal of Agriculture and Forestry, 5: 177-181.

Widjastuti, T.; Lengkey, H. A. A.W; Wiradimadja, R.; Herianti, D. 2011. Utilising waste product of Tuna (Thunnus atlanticus) fish silage and its implementation on the meat protein conversion of broiler. Lucrări Çtiintifice, Seria Zootehnie, 55: 83-87.

CHAPTER 3

PERFORMANCE, QUALITY AND SENSORY ANALYSIS OF EGGS LIGHT LAYERS FED ACIDIC PIRARUCU WASTE SILAGE MEAL

Performance, quality and sensory analysis of eggs light layers fed acidic pirarucu waste silage flour

SUMMARY

The aim of this study was to evaluate the performance, quality and sensory analysis of eggs from light commercial layers fed diets containing increasing levels of pirarucu waste acid silage flour. A total of 168 73-week-old Hisex White layers were used, distributed in a completely randomised experimental design with seven increasing levels of pirarucu waste acid silage flour (0; 0.5; 1.0; 1.5; 2.0; 2.5 and 3.0%) and four replicates of six birds each. The results of the treatments were subjected to polynomial regression analysis. The results obtained for feed consumption, laying percentage, feed conversion per kilo of feed per kilo of egg produced, feed conversion per kilo of feed per dozen eggs produced and egg mass of layers fed diets containing increasing levels of acid silage flour from pirarucu waste showed a significant difference ($P<0.05$) with a quadratic effect. Differences were also observed ($P<0.05$) in egg weight, yolk height, yolk pigmentation and egg flavour, with a quadratic effect. Based on the results found, it was concluded that pirarucu waste acid silage flour can be included up to a level of 2.5 per cent in the ration of light commercial layers without interfering with performance and egg quality.

KEYWORDS: alternative food, consumption, flavour, fish waste

1. INTRODUCTION

Today, poultry production, both broiler and layer, is recognised as an internationalised, uniform economic activity with no geographical borders. However, one of the obstacles faced by the sector in some Brazilian regions has been the need to import corn and soya, the main ingredients used to make feed, which often increase production costs due to regional distances.

In this context, the research carried out in the state on alternative foods that have the potential to be substituted or simply included in poultry feed is mainly aimed at minimising feed costs, which account for around 70% of total production costs. As a result, the use of agro-industry by-products in feed has reduced investment and offered animals a diet with high nutritional value, as well as being low-cost ingredients that are easily available in the regions (Enke et al. 2010).

One product that is currently being studied in Brazil to utilise fish waste is silage. Its production offers economic advantages, as it requires simple technology and is independent of scale (Hisano et al. 2012), as well as using material with low operating costs to transform waste materials into products of high nutritional quality that can minimise problems with environmental pollution, as well as serving as an ingredient in feed formulation.

Silage is a pasty product of high biological value resulting from the conservation of whole fish or their residual parts. Its principle consists of acidifying the pH through the addition of organic acids (Borghesi et al. 2008), which favour the autolytic enzymatic reactions present in the muscle, raising the levels of the fish.

of oligopeptides and free amino acids in the final product (Honorato et al. 2012).

The aim of this study was to evaluate the performance, quality and sensory analysis of eggs from light commercial layers fed diets containing increasing levels of acid silage flour produced from the residual biomass of pirarucu from the sustainable development reserve.

2. MATERIAL AND METHODS

The experiment was carried out in the Poultry Sector of the Faculty of Agricultural Sciences, located in the southern sector of the Federal University of Amazonas-UFAM, in Manaus-AM.

This study was registered (No. 040/2015) with the Ethics Committee for the Use of Animals (CEUA) at the Federal University of Amazonas in order to make the entire research feasible.

The experimental period was 84 days, from May to August 2016, divided into four periods of 21 days each. The birds were weighed (1.53±0.0025 kg) at the start of the experiment to standardise the plots and underwent a seven-day adaptation period to the experimental feed and facilities.

The experimental aviary used has a fibre cement roof, is 17.0m long, 3.5m wide and has a ceiling height of 3.20m, with a central aisle and a row of wire cages (1.0m x 0.45m x 0.45m) on each side, with internal partitions measuring 0.50m along the length, with linear trough-type feeders and niplle-type drinkers.

A total of 168 Hisex White layers aged 73 weeks were used, distributed in 28 cages in a completely randomised experimental design (DIC), where the treatments consisted of seven increasing levels of pirarucu residue acid silage meal 0; 0.5; 1.0; 1.5; 2.0; 2.5 and 3.0% in the feed and four replicates of six birds each. The experimental feed and water were provided ad libitum in each experimental unit. Eggs were collected twice a day, at 8am and 3pm, and their quantities recorded on the production sheet, according to the treatments and their respective repetitions.

During the experimental period, the birds received 16 hours of light (12 hours natural + 4 hours artificial), with artificial light supplied from 6pm to 8pm and from 4am to 6am. The average temperature and relative humidity were recorded twice a day (09:00 and 15:00) using a digital thermo-hygrometer positioned above the birds' cage in the aviary, with average results of 29.77±0.20°C and 79.85±0.01% respectively.

The acid silage was produced and dehydrated at the Fish Technology Laboratory of the Coordination of Technology and Innovation (COTI) of the National Institute for Amazonian Research (INPA). The waste from processing the pirarucu (ribs and vertebrae) was fragmented in a Multinox electric shredder. They were then divided into plastic buckets with a capacity of 15 litres, where each container received 10 kg of shredded mass. A mixture of formic and propionic acids were added to the containers in a 1:1 ratio, at 3% of the volume of the acid solution in relation to the weight of the waste under constant homogenisation, with the aim of acidifying the medium, lowering the pH to inhibit the proliferation of microorganisms and promoting protein hydrolysis, following the methodology proposed by Borghesi et al. (2008).

The containers were kept at room temperature (29 °C) for three days and turned every 24 hours to promote the incorporation of the acids into the silage mass. At the end of the 72 hours, to reduce humidity, the ensiled material was placed in aluminium trays and taken to a forced ventilation oven at 65 °C for 72 hours, during which time it was turned over to ensure uniform drying and obtain a quality dry product.

The chemical composition of acid silage and pirarucu waste was determined at the Food Chemistry and Physical Chemistry Laboratory, according to the methodology proposed by Silva and Queiroz (2012) and the minerals were digested in nitro-perchloric extract and the concentrations quantified by atomic absorption spectrophotometer (Ca) and by colorimetry (P) in the spectrophotometer using ammonium molybdate and ascorbic acid according to the methodology described by Sarruge and Haag (1974) at the Soil and Plant Chemistry Thematic Laboratory of the National Amazon Research Institute (INPA). After determining the chemical composition (Table 1), the product was packaged and sent to the poultry sector at UFAM, where it was ground to obtain acid silage flour from pirarucu waste.

Table 1. Chemical composition of pirarucu waste acid silage flour*.

Components	Composition
Dry Matter, %	84,16
Crude Protein, %	40,05
Ether extract, %	26,81
Mineral Matter, %	9,31
Total Carbohydrates, %	7,97
Calcium, g/kg	65,16
Phosphorus, g/kg	22,90
Metabolizable energy, kcal/kg	3253,01**

* Averages taken from three repetitions.
**Determined using the calculation method for apparent metabolisable energy as described by Rostagno et al. (2011), where the value in kcal kg^{-1} was obtained.

The experimental diets were formulated according to the nutritional requirements of the layers using the computer programme Supercrac (2004) using the values provided by Rostagno et al. (2011) and the inclusion of acidic pirarucu waste silage respecting the percentage inclusion levels in the rations (Table 2).

Table 2. Formulations of the experimental rations with the levels of inclusion of acidic pirarucu waste silage in the diet of light layers.

Ingredients	Levels of acidic pirarucu waste silage per treatment						
	0,0%	0,5 %	1,0 %	1,5 %	2,0 %	2,5 %	3,0 %
Maize (7.88%)	62,255	62, 170	62,087	62,005	61,922	61,839	61,756

F. soya (46%)	25,871	25,527	25,173	24,819	24,466	24,112	23,758
Sour silage	0,000	0,500	1,000	1,500	2,000	2,500	3,000
Calcitic limestone	9,239	9,167	9,095	9,023	8,950	8,878	8,806
Bicalcium phosphate	1,693	1,696	1,700	1,704	1,708	1,712	1,716
Vitamin and mineral supplement[1]	0,500	0,500	0,500	0,500	0,500	0,500	0,500
DL - Methionine (99%)	0,092	0,089	0,094	0,099	0,104	0,109	0,113
Salt	0,350	0,350	0,350	0,350	0,350	0,350	0,350
Total	100,00	100,00	100,00	100,00	100,00	100,00	100,00
Nutrient	Nutritional levels						
E.M, kcal^{-1}/kg	2692,6	2686,8	2671,3	2660,8	2650,2	2639,7	2629,2
Crude protein, %	17,000	17,000	17,000	17,000	17,000	17,000	17,000
Methionine + Cystine, %	0,627	0,620	0,620	0,620	0,620	0,620	0,620
Methionine, %	0,360	0,361	0,369	0,378	0,387	0,395	0,404
Calcium, %	4,000	4,000	4,000	4,000	4,000	4,000	4,000
Available Phosphorus, %	0,400	0,400	0,400	0,400	0,400	0,400	0,400
Sodium, %	0,156	0,156	0,156	0,156	0,156	0,156	0,151

[1] Guarantee levels per kilogram of product: Vitamin A 2,000,000 IU, Vitamin D3 400,000 IU, Vitamin E 2,400 mg, Vitamin K3 400 mg, Vitamin B1 100 mg, Vitamin B2 760 mg, Vitamin B6 100 mg, Vitamin B12 2.400 mcg, Niacin 5,000 mg, Calcium Pantothenate 2000 mg, Folic Acid 50 mg, Cocciostatic 12,000 mg, Choline 50,000 mg, Copper 1,200 mg, Iron 6,000 mg, Manganese 14,000 mg, Zinc 10,000 mg, Iodine 100 mg. Selenium 40 mg. Vehicle Q.S.P. 1.000 g.

2.1 Parameters Studied in Zootechnical Performance

- **Feed consumption**

Feed consumption was determined by the ratio between the total amount of feed consumed and the number of birds in a 21-day period, based on the amount of feed offered during the cycle, minus the surplus at the end of each cycle. To calculate feed consumption, the number of birds killed during the cycle was taken into account.

- **Feed conversion (kg feed/kg egg mass)**

Through data on the total feed consumed and the total egg mass produced.

- **Feed conversion (kg/dz)**

Through data on the total amount of feed consumed and the total number of dozen eggs produced.

- **Egg production (%)**

It was determined by taking the total number of eggs produced and the total number of possible eggs that could be produced, multiplied by one hundred.

- **Egg mass (g)**

Egg mass was obtained by calculating the product of egg weight and the number of eggs produced divided by one hundred.

- **Mortality (%)**

It was determined by multiplying the number of dead birds by the total number of birds at the beginning of the plot, multiplied by one hundred.

On the last two days of each period, four eggs were randomly collected from each plot to measure egg quality. Before being evaluated, the eggs were stored for one hour until their temperature was equalised with the laboratory temperature of 25 °C.

2.1 Egg Quality Parameters

- **Egg weight (g)**

The eggs were separated by repetition and weighed on an analytical balance. The egg mass was obtained by calculating the quotient between egg weight and egg production multiplied by one hundred.

- **Specific Gravity (g/ml^3)**

Immediately after collection, the whole eggs were placed in iron baskets and submerged in plastic buckets containing different solutions of sodium chloride (NaCl), from the lowest to the highest concentration, with a density ranging from 1,075 to 1,100 g/cm^3, with an interval of 0.005 between them. The eggs were placed in a bucket of pure water before being transferred to the salt solutions. The eggs were removed as they floated to the surface and their respective values recorded.

- **Albumen and yolk height (mm)**

After weighing the eggs, they were broken on a flat glass plate to measure the height of the albumen and yolk. This procedure consists of measuring the median area between the outer edge of the albumen and the yolk using an electronic caliper.

- **Yolk pigmentation**

This variable was found using a ROCHE colourimetric fan with a numerical scale from 1 to 15.

- **Percentage of Albumen and Yolk (%)**

To analyse the percentage of albumen and yolk, a manual albumen and yolk separator was used. The albumen was placed in a beaker and weighed on an analytical balance. The same procedure was used to measure the weight of the yolk. The values in grams were recorded. For the percentage of both, the weights were divided by the weight of the egg and multiplied by one hundred, giving the values in percentage.

- **Bark percentage (%)**

The eggshell percentage was obtained after the shells had spent 2 hours in a hot chamber to dry. After this period, an analytical balance was used to weigh the shells and record the values in grams. For the percentage, the weights were divided by the weight of the egg and multiplied by one hundred, giving the percentage values.

- **Bark thickness (jim)**

The dried shells were used for this determination. Readings were taken using a micrometer at three points: the basal, equatorial and apical regions of the egg. The average was taken in micrometres and the values recorded.

- **Haugh Unit**

To determine the Haugh unit, the mathematical formula proposed by Haugh (1937) was used, described as: $UH = 100 \times \log (H + 7.57 - 1.7 \times W^{0.37})$, where: H = height of dense albumen (mm) and W = egg weight (g).

2.2 Sensory analysis of the eggs

The sensory evaluation was carried out according to the methodology proposed by Dutcosky (2007), using 46 untrained judges of both sexes. Each participant was given an evaluation form containing a nine-point Structured Hedonic Scale (9-very much liked; 8-very much liked; 7-moderately liked; 6-slightly liked; 5-neither liked nor disliked; 4-slightly disliked; 3-moderately disliked; 2-very much disliked; 1-very much disliked), for the following attributes aroma, appearance, colour and taste. Forty-two experimental eggs were used (6 eggs from each treatment). These eggs were cooked in boiling water (10 litres) for 10 minutes after boiling, at a temperature of 72°C. The cooked eggs were then removed and left to cool at room temperature. After this procedure, the judges were given samples of each treatment.

2.3 Statistical Analysis

The data collected was subjected to statistical analysis using the Statistical Analysis System - SAS (2008) computer programme and the estimates of the treatments were subjected to regression analysis. The Tuckey test was used to compare the means of the treatments at a 0.05% probability level.

RESULTS AND DISCUSSION

The results obtained for layer performance are shown in Table 3. There were significant differences ($P<0.05$) in the results of feed consumption, laying percentage, feed conversion (kg/kg and kg/dz) and egg mass of layers fed rations containing increasing levels of acidic pirarucu waste silage flour.

Feed consumption showed a quadratic effect ($Y= 108.67 + 1.7704x - 0.3122x2$, $R^2 = 0.82$) at the level of inclusion of acid silage meal from pirarucu waste, with an estimated consumption of 111.18 g/bird/day for layers fed diets containing 2.38% acid silage meal compared to the other levels used. The test of means showed that feed consumption with the last level of inclusion of acid silage flour in the feed was lower than the other levels and the control treatment.

Table 3. Feed consumption (FC), laying percentage (LOS), egg mass (OM), feed conversion per kilo of feed per kilo of egg produced (FC, kg/kg) and feed conversion per kilo of feed per dozen eggs produced (FC, kg/dz).

Variables	Pirarucu chemical silage inclusion levels (%)							P Value	R^2	CV, %
	0,00	0,50	1,00	1,50	2,00	2,50	3,00			
CR, g/bird/day	110,42	110,84	110,81	110,52	109,24	109,89	104,78*	0,01	0,82	0,46
PERCP, %	89,98	93,90	88,09	83,93	89,53	89,38	66,86*	0,01	0,62	5,14
CA, kg/kg	1,56	1,60	1,62	1,63	1,59	1,58	2,01*	0,01	0,61	5,06
CA, kg/dz	2,92	2,81	3,00	3,14	2,91	2,93	3,74*	0,01	0,58	5,63
MO, g	70,50	68,98	68,34	67,66	68,57	69,45	52,42*	0,01	0,64	4,43

CV - Coefficient of variation. P Value - Coefficient of Probability. R2 - Coefficient of determination. *Significant difference ($P<0.05$) by Tukey test.

This reduction in feed consumption by layers fed the last inclusion level is related to the energy content of the acidic pirarucu waste silage meal, since excess fat in the diet leads to satiety in the animal and causes changes in consumption levels as a controlling factor. According to Carioca et al. (2010), the more energy the feed has, the less it is consumed by the layers. This reduction in feed intake is caused by the presence of lipid solutions in the duodenum that stimulate the release of the hormone cholecystokinin (CCK), which acts on the enterogastric reflex, slowing down the rate at which feed passes through the gastrointestinal tract (Brunelli et al. 2010; Pinheiro et al. 2012; Torres and Drehe 2015).

However, modern layers are able to adjust their feed intake according to their energy needs (Gunawardana et al. 2008; Ribeiro et al. 2014). In the present study, the reduction in feed consumption was recorded in the second evaluation cycle and lasted until the first week of the third. From this point onwards, the layers returned to normal feed consumption .

The feed consumption results found in this study disagree with the work carried out by Enke et al. (2010), who worked with acid silage flour from fish waste, added to defatted rice bran in the feed of Japanese quails (Coturnix coturnix japonica), in which they found no significant difference between the diets tested.

It is worth noting that studies carried out to evaluate broilers fed with increasing levels of fish waste silage inclusion in the diet have obtained different results, such as those observed by Valenzuela et al. (2015) who found an increase in consumption of 90-128g/bird up to the level of 15% inclusion of acidic salmon silage, and Ramirez-Ramirez et al. (2016) who found that inclusion levels of up to 30% of this in broiler diets did not influence feed consumption.

The laying percentage showed a quadratic effect at the increasing levels of acidic pirarucu waste silage meal, where the derivation of the regression equation ($Y = 84.892 + 5.2881x - 1.0045x^2$, $R^2 = 0.62$) indicated that the best production index (91.85) was obtained at the 2.63% acidic silage inclusion level. Consequently, when evaluating egg mass, a quadratic effect was also observed ($Y = 64.611 + 4.4612x - 0.7946x^2$, R2 = 0.64), with the maximum egg mass of 70.87g being obtained at the 2.80% inclusion level of acid silage from pirarucu waste in the feed.

It was observed that layers fed lower levels of acid silage meal from pirarucu waste in their feed had better egg production and mass compared to those that consumed the last level of acid silage meal. The response to this decrease in feed intake led to a reduction in egg production and mass. According to Perez-Bonilla et al. (2012), changes in diet composition can result in an increase in body weight gain and a decrease in egg mass production.

Feed conversion (kg/kg) increased quadratically ($Y = 1.7239 - 0.1173x + 0.0203x2$, $R^2 = 0.61$), with a maximum of 1.55 kg/kg at the 2.88% inclusion level of acidic pirarucu waste silage meal in the feed. Corroborating this result, feed conversion (kg of feed/dozen eggs) also showed a quadratic effect ($Y = 3.1289 - 0.1956x + 0.0361x^2$, $R2 = 0.58$), with the best conversion (2.86 kg/dozen eggs) being obtained at the 2.70% inclusion level. These results were similar to those observed by Hanna et al. (2013) studying increasing levels of copaiba oil in the feed of 88-week-old Hisex White layers, who also found the same effects with increasing levels in the feed.

Feed conversions (kg/kg and kg/dozen) showed worse results at the highest level of inclusion of pirarucu waste acid silage flour in the feed, verifying a direct relationship with consumption, since when there was a reduction in feed consumption by the layers, there was an increase in feed conversions and consequently a decrease in egg production and mass.

Table 4 shows the egg quality results. Egg weight showed a quadratic effect ($Y = 59.226 + 3.2783x - 0.5817x2$, $R2 = 0.58$) where, by deriving the function, it was possible to estimate a maximum weight of 63.84g by including 2.81% of acid silage meal from pirarucu waste in the feed.

Table 4. Egg weight (OP), albumen percentage (PERA), yolk percentage (PERG), shell percentage (PERC), albumen height (AA), yolk height (AG), shell thickness (EC), specific gravity (SG), Haugh unit (HU) and yolk pigmentation (PG) of eggs from light layers.

	Piraracu chemical silage inclusion levels (%)							P Value		CV, %
Variables	0,0	0,5	1,0	1,5	2,0	2,5	3,0		R^2	
PO, g	63,85	62,23	61,67	61,21	62,77	63,19	50,01*	0,01	0,58	4,31
PERA, %	54,59	54,66	56,72	53,69	53,63	56,02	52,91	0,07	-	3,28
PERG, %	26,42	27,05	27,18	27,88	27,89	27,87	26,85	0,31	-	3,80
PERC, %	9,43	9,55	9,60	10,01	9,20	9,39	9,73	0,09	-	3,77
AA, mm	7,73	7,21	7,68	7,86	7,50	7,77	7,49	0,29	-	5,04
GA, mm	17,72	17,43	18,06	18,77	19,42	18,67	17,59	0,05	0,57	5,15
EC, gm	33,77	35,90	35,17	35,35	35,09	35,02	36,26	0,32	-	4,00
GE, g/mL	1084,92	1102,26	1084,22	1087,11	1083,05	1083,91	1133,54	0,43	-	3,34
UH	85,55	84,15	86,90	87,63	85,49	86,92	85,87	0,43	-	2,68
PG	4,79	5,23	5,17	5,36*	5,29	5,26	4,73*	0,01	0,84	5,04

CV - Coefficient of variation. Quadratic effect ($P<0.05$). P Value - Coefficient of Probability. R2 - Coefficient of determination. * significant difference ($P<0.05$) by Tukey test.

The mean comparison test showed that egg weight decreased as the level of 3.0 per cent inclusion of acid silage meal from pirarucu waste in the feed increased, and that this difference is related to the reflection of consumption caused by the higher level of inclusion. These results disagree with those observed by Jonsson et al. (2011) who found no significant difference in the egg weight of layers fed 3.5% mussel meal in their diet.

According to Well et al. (2010) changes in egg weight reduction through diets containing increasing levels of fishmeal or fish oil are possibly associated with omega-3 in the diet, and changes in lipid metabolism in birds. These changes can lead to a reduction in the surrounding triglycerides in the blood plasma and have a limiting effect on the availability of lipids for yolk formation (Kamely et al. 2016).

Yolk height showed a quadratic effect ($Y= 16.178 + 1.1683x - 0.1306x2$, $R^2 = 0.57$), where it was possible to estimate the maximum peak of 18.43 mm at the 2.83% inclusion level of acid silage meal from pirarucu waste in the feed. These results are related to the omega-3 and omega-6 content present in pirarucu waste (Sherr et al. 2014), which through the acid silage process at levels of inclusion in the feed contributed to the increase in yolk height.

According to the studies carried out to test the inclusion levels of fishmeal and fish oil in the diet of layers and quails, Al-Daraji et al. (2011); Alemayehu et al. (2015); Amao et al. (2010) observed a significant increase in yolk height with the inclusion levels of these products in the feed.

A quadratic effect was observed for egg pigmentation ($Y = 4.3817 + 0.494x - 0.0618x^2$, R2 = 0.84), with a maximum colour value (5.31) at the 2.99% inclusion level of acidic pirarucu waste silage flour. Parede et al. (2009) found that inclusion levels of 3.5% mussel meal in layer feed influenced yolk pigmentation.

The colour of the egg yolk is considered an important factor in determining the product for consumers. Thus, the 1.5% inclusion level resulted in more intense yolk pigmentation in the Roche score and less intense colouration was recorded for the last inclusion level. This result is attributed to the fat content present in the pirarucu ribs, used to make the acid silage, which contributed to the changes in yolk pigmentation in the different treatments.

The colour of the yolk depends on the type of ingredient used in the feed and whether it contains the pigments needed to develop colour. In this context, Saleh (2013) observed that the inclusion of fish oil contributed to the darkening of the yolk at levels of 5% in the diet of layers and Jonsson et al. (2011) observed that layers fed 7% mussel meal in the feed produced stronger yolk colour.

The results for albumen percentage, yolk percentage, shell percentage, albumen height, shell thickness, specific gravity and Haugh unit were not influenced (P>0.05) by the inclusion levels of acidic pirarucu waste silage meal in the diet of 73-week-old light layers. Results similar to these were observed by Well et al. (2010) who found no significant difference at inclusion levels of 3.5 and 7.0 per cent mussel meal for most of the egg quality studied.

Table 5 shows the data obtained from the sensory analysis of eggs from layers fed increasing levels of acidic pirarucu waste silage flour. Of the attributes assessed in the sensory analysis, there were no significant differences in the variables analysed, and the regression analysis indicated a quadratic effect ($Y = 7.333 - 0.249x + 0.0308x^2$, $R^{(2)} = 0.60$), with an estimate of (6.89) at the 4.04% inclusion level, indicating that increasing the inclusion of pirarucu acid silage flour in the feed does not influence or alter the taste of the eggs.The evaluators gave scores between 6.33 and 7.58 indicating that they liked it slightly and liked it moderately. These results agree with those reported by Yi et al. (2014) who found no significant differences in flavour between eggs from layers fed salmon meal compared to the control diet.

Table 5. Aroma (ARO), colour (COR), taste (SAB) and appearance (APA) of eggs from layers fed diets containing increasing levels of chemical pirarucu silage (%).

Variables	Piraruсu chemical silage inclusion levels (%)							P Value	R^2	CV, %
	0,0	0,5	1,0	1,5	2,0	2,5	3,0			
ARO	6,50	6,50	5,91	6,41	6,91	6,58	6,41	0,87	-	24,84
COLOUR	6,91	5,91	6,00	5,91	6,83	6,66	6,41	0,54	-	25,88
SAB	7,41	6,83	6,33	6,66	7,25	7,58	6,58	0,05	0,60	20,86
APA	7,16	7,33	7,08	6,50	7,16	7,66	7,25	0,61	-	19,66

CV - Coefficient of variation. Quadratic effect (P<0.05). P Value - Coefficient of Probability. R2 - Coefficient of determination. * significant difference (P<0.05) by Tukey test.

Seibel et al. (2010) found that including the solid fraction of acidic fish silage with (3.3; 6.6; and 9.9%) and without (5 to 10%) antioxidants in the feed provided less sensory changes than commercial semi-purified fish oil. According to the authors, when antioxidants are added to silage, it has no effect on the odour and taste of the eggs. In the present study, no unpleasant fishy flavour was detected, even without the addition of antioxidants, in eggs produced with the levels of inclusion of acidic pirarucu waste silage flour in the feed.

Alemayehu et al. (2015) observed that hens fed diets containing 10-15% fish waste meal showed a pronounced fishy flavour in the eggs compared to those fed a 5% inclusion, which showed no difference to the control diet. Fish products contain long-chain fatty acids (Jonsson and Elwinger 2009), which can develop an unpleasant flavour in eggs as a result of fatty acid oxidation (Jonsson et al 2011).

The aroma, colour and appearance of the eggs were not affected (P>0.05) by the inclusion levels of pirarucu acid silage meal. Similar results were observed by Carvalho et al. (2010) who found no significant difference in the parameters assessed.

3. CONCLUSION

The results obtained in this study show that pirarucu waste acid silage flour can be used as an alternative feed in rations for light commercial layers. Up to the 2.5 per cent inclusion level, there was no negative interference in performance or egg quality.

4. REFERENCES

Al-Daraji. H. J. Al-Mashadani, H.A.; Al-Hayani, W.K.; Mirza, H. A.; Al-Hassani, A.S. 2011. Influence of Source of Oil Added to Diet on Egg Quality Traits of Laying Quail. International Journal of Poutry Science, 10: 130-136.

Alemayehu, Y.; Urge, M.; Getu, A. 2015. Effects of Levels of Inclusion of Locally Processed Fish Waste Meal in the Diets of White Leghorn Layers on Performance Parameters, Hatchability, Economics, Egg Production and Egg Quality. Iranian Journal of Applied Animal Science, 5: 689-698.

Amao, O. A.; Oladunjoye, I. O; Togun, V. A.; Olubajo, K.; Oyaniyi, O. 2010. Effect of Westwood (Cirina forda) Larva Meal on the Laying Performance and Egg Characteristics of Layng Hen in a Tropical Environment. International Journal of Poultry Science, 9: 450-454.

Borghesi, R.; Portz. L; Oetterer, M.; Cyrino, J. E. P. 2008. Apparent digestibility coefficient of protein and amino acids of acid, biological and enzymatic silage for Nile tilapia (Oreochromis niloticus|). Aquaculture Nutrition, 14: 242-248.

Brunelli, S. R.; Pinheiro, J. W.; Fonseca, N. A. N.; Oba, A.; Silva, C. A. 2010. Defatted maize germ meal in the diet of commercial layers from 28 to 44 weeks of age. Revista Brasileira de Zootecnia, 39: 1068-1073.

Brunelli, S. R.; Pinheiro, J. W.; Fonseca, N. A. N.; Silva, C. A. 2012. Effect of different levels of defatted maize germ meal in diets supplemented with phytase for commercial layers. Agricultural Sciences, 33: 1991-2000.

Carioca, S.T.; Guimarães Cruz, F.G.; Matos, P.G.J; Maquiné, L.C.; Chagas, E.O.; Oliveira, S.C.; Santos Filho, A.B. 2010. Influence of energy and protein levels in rations for light layers in Manaus. Archivos da Zootecnia, 59: 455-458.

Carvalho, P. R. de; Pita, M. C. G.; Piber Neto, E.; Mendonça Júnior, C. X. 2010. Influence of the Supplementary Addition of Rich Sources of PUFAS n-3 of Marine Origin in the Diet on the Organoleptic Properties of Chicken Eggs. Scientific Journal of Animal Production. 12: 236-239.

Dutcosky, S. D. Sensory analysis of food. Curitiba: Champagnat, 2007. 239 p.

Enke, D. B. S; Tabeleão, V.; Rocha, C. B.; Rutz, F.; Soares, L. A. de S. 2010. Effect of the Inclusion of Fish Silage Flour Added to Defatted Rice Bran in the Diet of Japanese Quails (Coturnix Coturnix Japonica). Brazilian Journal of Animal Hygiene and Health, 4: 01-14.

Gunawardana, P.; Roland Sr. D. A.; Bryant, M. M. 2008. Effect of Energy and Protein on Performance, Egg Components, Egg Solids, Egg Quality, and Profits in Molted Hy- Line W-36 Hens. The Journal of Applied Poultry Research, 17: 432-439.

Haugh, R. R. 1937. The Haugh unit for measuring egg quality. United States Egg Poultry Magazine, v. 43, p. 552-555.

Hanna, A. C. S.; Cruz, F. G.; Rufino, J. P. F; Tanaka, E. de S.; Chagas, E. O.; Melo, J. B. dos S. 2013. Bioefficacy of the Copaiba Oil (Copaifera sp.) In Diets of Laying Hens in the Second Production Cycle in Humid Tropical Climate. International Journal of Poultry Science, 12: 647-652.

Hisano, H.; Ishikawa, M. M.; Portz, L. 2012. Production of acid silage from surubim (Pseudoplatystoma sp.) viscera and evaluation of digestibility for nile tilapia. Revista Brasileira de Saúde e Produção Animal, Salvador, 13: 872-879.

Honorato, C.; Frizzas, G. O.; Carneiro, D. J. 2012. Digestibility of fish silage with different storage times for feeding pacu (piaractus mesopotamicus). Trials and Science: Biological, Agricultural and Health Sciences, 16: 8595.

Jonsson, L.; Wall, H.; Tauson, R. 2011. Production and egg quality in layers fed organic diets with mussel meal. Animal, 5: 387-393.

Jonsson, L.; Elwinger, K. 2009. Mussel meal as a replacement for fish meal in feeds for organic poultry - a pilot short term study. Acta Agriculturae Scandinavica - Section A, Animal Science, 59: 22-27.

Kamely, M.; Karimi Torshizi, M. A.; Khosravinia, H. 2016. Omega-3 Enrichment of Quail Eggs: Age, Fish Oil, and Savory Essential Oil. Journal of Agricultural Science and Technology, 18: 347-359.

Parede, H.; Jonsson, L.; Johansson, L. 2009. Effects on egg quality traits of genotype and diets with mussel meal or wheat-distillers dried grains with solubles. Poultry Science, 89:745-751.

Pérez-Bonilla, A.; Novoa, S.; García, J.; Mohiti-Asli, M.; Frikha, M.; Mateos, G. G. 2012. Effects of energy concentration of the diet on productive performance and egg quality of brown egg-laying hens differing in initial body weight. Poultry Science, 91: 3156-3166.

Pinheiro, J. W.; Fonseca, N. A. N.; Bridi, A. M.; Silva, C. A.; Oba, A.; Medeiros, L. G.; Oliveira, M. L. L. 2012. Performance and egg quality of light commercial layers fed diets containing turnip cake (Raphanus sativus). Agricultural Sciences, 33: 1555-1564.

Rahman, M.; Koh, K. 2016. Effects of Formic Acid-Treated Shrimp Meal on Growth Performance and Nutrient Digestibility in Broilers. Japan Poultry Science Association, 53: 208-212.

Ramírez-Ramírez, J. C.; Espain, J. I. I.; Leyva, R. G.; Ulloa, J. A.; Ulloa, P. R. 2016. Use of biological fish silage in broilers feed: Effect on growth performance and meat quality. Journal of Animal & Plant Sciences, 27: 4293-4304.

Ribeiro, P.; Matos, Jr J. B.; Lara L. J. C; Araújo L.F; Albuquerque R.; Baião N. C. 2014. Effect of Dietary Energy Concentration on Performance Parameters and Egg Quality of White Leghorn Laying Hens. Brazilian Journal of Poultry Science, 16: 381388.

Rostagno, H.S.; Albino, L. F. T.; Donzele, J. L.; Gomes, P. C.; Oliveira, R. F. de; Lopes, D. C.; Ferreira, A. S.; Barreto, S. L. de T. Ricardo Frederico Euclides. Brazilian tables for poultry and pigs: feed composition and nutritional requirements. Viçosa: Federal University of Viçosa, 2011. 252p.

Sarruge, J. R. & Haag, H. P. Chemical analyses in plants. Piracicaba: USP-ESALQ, 1974. 65p.

Saleh, A. A. 2013. Effects of fish oil on the production performances, polyunsaturated fatty acids and cholesterol levels of yolk in hens. Emirates Journal Food Agriculture, 25: 605-612.

Seibel, N. F.; Schoffen, D. B.; Queiroz, M. I.; Souza-Soares, L. A. 2010. Sensory characterisation of quail eggs fed modified diets. Food Science and Technology, 30: 884-889.

Silva, D.J.; Queiroz, A.C. Food analysis: chemical and biological methods. Viçosa: Federal University of Viçosa, 2012.

Supercrac. Minimum cost ration. Version 1.02, for Windows: TD Software, 2004.

Scherr, C. Gagliardi, A. C. M.; Miname, M. H.; Santos, R. D. 2014. Fatty Acid and Cholesterol Concentrations in Usually Consumed Fish in Brazil. Arquivos Brasileiros de Cardiologia, 104: 152-158.

Statistical Analysis System - SAS. SAS/STAT Software Version 9.2. Cary: SAS Institute Inc., 2008.

Torres, R. de N. S.; Dreher, A. 2015. Lipid sources in layer diets: Production and egg quality. Nutritime Electronic Journal, 12: 3952-3963.

Valenzuela, C.; Carvallob, F; Moralesa, M. S.; Reyesa, P. 2015. Efecto del uso de ensilado seco de salmón en dietas de pollos broiler sobre parámetros productivos y calidad sensorial de la carne. Archivos de medicina veterinária, 47: 53-59.

Wall, H.; Jonsson, L.; Johansson, L. 2010. Effects on egg quality traits of genotype and diets with mussel meal or wheat-distillers dried grains with solubles. Poultry Science, 89: 745-751.

Yi, H.; Hwang, K. T.; Regenstein, J. M.; Shin, S. W. 2014. Fatty Acid Composition and Sensory Characteristics of Eggs Obtained from Hens Fed Flaxseed Oil, Dried Whitebait and/or Fructo-oligosaccharide. Asian-Australasian Journal of Animal Sciences. 27: 1026-1034.

FSC
www.fsc.org
MIX
Papier | Fördert gute Waldnutzung
FSC® C083411

Kolibri 360 GmbH
Ferdinand-Jühlke-Straße 7
99095 Erfurt, Deutschland
produktsicherheit@kolibri360.de

Druck:
CPI Druckdienstleistungen GmbH
im Auftrag der
Kolibri 360 GmbH
Ferdinand-Jühlke-Str. 7
99095 Erfurt